P9-CKW-920

# The Mutagenicity of Pesticides
## Concepts and Evaluation

The MIT Press
Cambridge, Massachusetts, and London, England

# The Mutagenicity of Pesticides
## Concepts and Evaluation

Samuel S. Epstein and Marvin S. Legator

with a foreword by Joshua Lederberg

This book was designed by
the MIT Press Design Department.
It was set in Univers 685/689
Printed on 60 lb. P&S Offset
by Halliday Lithograph Corp.
and bound by The Colonial Press Inc.
in the United States of America



ISBN (hard cover) 0 262 05008 0

Library of Congress catalog card number: 70–130274

To Alexander Hollaender
for his pioneering work in chemical mutagenesis

## List of Contributors and Participants


Dr. S. Abrahamson
The University of Wisconsin
Madison, Wisconsin

Dr. P. J. Bottino
Brookhaven National Laboratory
Upton, Long Island, New York

Dr. J. Crow
The University of Wisconsin
Madison, Wisconsin

Dr. R. Donnelly
Food and Drug Administration
and George Washington University,
Washington, D.C.

Dr. S. S. Epstein
Harvard Medical School and
Children's Cancer Research
Foundation, Inc.,
Boston, Massachusetts

Dr. E. Freese
National Institute of Neurological
Diseases and Blindness
Bethesda, Maryland

Mr. I. Gerring
Division of Research Grants
National Institutes of Health
Bethesda, Maryland

Dr. A. C. Kolbye
Bureau of Food and Nutrition
Food and Drug Administration
Washington, D.C.

Dr. J. Lederberg
Stanford University School of Medicine
Palo Alto, California

Dr. M. S. Legator
Food and Drug Administration
and George Washington University,
Washington, D.C.

Dr. H. V. Malling
Oak Ridge National Laboratory
Oak Ridge, Tennessee

Dr. R. S. McCutcheon
Division of Research Grants
National Institutes of Health
Bethesda, Maryland

Dr. J. Neumeyer
Northeastern University
Boston, Massachusetts

Dr. W. Nichols
Institute of Medical Research
Camden, New Jersey

Dr. L. A. Schaierer
Brookhaven National Laboratory
Upton, Long Island, New York

Dr. A. H. Sparrow
Brookhaven National Laboratory
Upton, Long Island, New York

Dr. J. S. Wassom
Oak Ridge National Laboratory
Oak Ridge, Tennessee

Dr. C. Wilkinson
Cornell University
Ithaca, New York

# Foreword

Socially responsible scientists should be concerned about the potential hazards of chemically induced mutation for at least three reasons. The most important is also the most remote in the scale of time: the human nature that defines our posterity is energized by our cultural tradition; but it is also bounded by the integrity of the genetic information of which each generation is the vessel.

Second, genetic impairments already account for a very large part of our existing burden of disease and premature death. If we give proper weight to the genetic component of many common diseases which have a more complex etiology than the textbook examples of Mendelian defects, we can calculate that at least 25 percent of our health burden is of genetic origin. This figure is a very conservative estimate in view of the genetic component of such griefs as schizophrenia, diabetes, atherosclerosis, mental retardation, early senility, and many congenital malformations. In fact, the genetic factor in disease is bound to increase to an even larger proportion, for as we deal with infectious disease and other environmental insults, the genetic legacy of the species will compete only with traumatic accidents as the major factor in health.

Finally, experimental evidence of mutagenic capability should be a danger signal that a compound may also be capable of other, somatic hazards through its action on DNA and other cellular constituents. For example, the demonstration of chromosome breakage in cultured cells exposed to cyclohexylamine was the (administratively neglected) forerunner of the eventual inculpation of the parent compound, cyclamate, as a cancer hazard.[3]

Our existing genetic load is a summation of three kinds of process: the historical accumulation of recessive gene mutations reappearing from time to time as their heterozygous carriers chance to mate; the immediate manifestation of dominant mutations and chromosome anomalies, which are only rarely propagated; and the paradoxical segregational load, where deleterious recessives had been stabilized within the population through some present, or more often historical, advantage of the heterozygotes.

Any assessment of the social and personal costs of mutation must take account both of absolute and of relative measures. (And of course we must use the same perspective in weighing the social and personal benefits claimed for a given environmental additive.) A 10 percent increase in the existing, "spontaneous" mutation rate is, in effect, the standard that has been adopted as the "maximum acceptable" level of public exposure to radiation by responsible regulatory bodies.[4] This level can be defended on the argument that we neglect to take a number of measures that could probably improve the mutation index to a comparable degree. It can be attacked by reciting the absolute level of eventual biological injury that might come from public exposure at such a level, were this in fact to occur from the proliferation of nuclear power plants and unregulated weapons tests.

A rational approach to the assessment of chemical hazards likewise calls for a detailed *quantitative* examination of risk, if only to expose the policy assumptions to the same degree of public understanding and debate as pertains to atomic energy. In both sets of situations, we are plagued by serious uncertainties about the numerical estimates for induced mutation in man, perhaps much worse for any chemical than for radiation. Given the problem of the number of new, suspicious compounds now pervading the environment, we face a formidable task in putting our genetic house in order.

For many purposes, we could dispose with a quantitative analysis, simply because most of our assays are so *in*sensitive that a compound that scores positively as a mutagen must have a portentous effect in any but the most peculiar circumstances. Nevertheless, the labeling of compounds as "mutagenic" or "nonmutagenic" may, particularly if we pursue the development of assays for sensitivity rather than selectivity, be regarded as simplistic, both in quantitative and in qualitative terms. However, when there are murky doubts about an issue as important as mutational damage, they must be resolved in favor of the species.

The sobriety with which we face the task of setting up rational criteria for decision must be robust enough to withstand some inevitable ridicule. The very first compound to be reported[1] in the published literature as mutagenic was *allyl isothiocyanate*, or mustard oil, well known as a natural constituent of horseradish and other widely used condiments. (This should not be confused with mustard gas, research on which was classified during World War II, and which was later revealed to be a mutagen of uncommon potency.) Whether allyl isothiocyanate is mutagenic in mammals is not known, and it is too soon to condemn a foodstuff having such a venerable tradition. This is a concession to that tradition, rather than to any factual assurance that mustard-eaters have a mutation rate demonstrably unaltered by their diet. Further research may allay or confirm these suspicions; if they are confirmed, it would not be the first time that a common dietary article was found to be harmful to some consumers despite centuries of common use (cf. the role of wheat gluten, only recently understood in the etiology of celiac disease.)

This kind of diffidence in the face of custom and uncertainty does not extend to efforts to promulgate a compound like mustard oil for uses lacking an ancient precedent. It has, for example, been advocated as a way of denaturing hobby glues to deter sniffing, an admirable social purpose if it reached the roots of the problem. The FDA's approval of mustard oil as an obviously safe additive is the entire argument against unwonted side effects. This must be balanced against a long-buried observation of skin tumors in mice painted with mustard oil;[7] which is also another example of the predictive value of mutagenicity for carcinogenicity. Theoretically, isothiocyanates can be expected to function as alkylating agents.

Mustard oil is not a pesticide (except, perhaps, as a natural one evolved by cruciferous plants); why mention it here? Mainly because, little as we know about it, more is known about mustard oil than about most pesticides. Furthermore, most of our concerns about pesticides are prototypes of our general concerns about environmental additives. We should admit, at the outset, that pesticides must be differentiated from food additives in the dilemmas that they raise about the cost-benefit equation. (Some critics believe, however, that pesticides are overpromoted to the point where economic benefits would be left intact or even improved if they were used more selectively; and the environmental load could then be cut by a factor of 10 or 100.)

It will encourage a fuller exposure of the magnitude and incidence of economic benefits of potentially hazardous chemicals if we do suggest some quantitative standards of acceptable mutagenicity. I believe that the present standards for population exposure to radiation should and will (at least de facto) be made more stringent, to about 1 percent of the spontaneous rate, and that this is a reasonable standard for the maximum tolerable mutagenic effect of any environmental chemical (better, for them in aggregate).

Accepting, for present argument, the formal arguments of the UN advisory group,[8] I translate this standard into a rate of about one recessive mutation per 1,000 gametes ($10^{-7}$ per nominal locus) per generation of typical exposure. Dominant mutations and chromosome aberrations may deserve even more stringent scrutiny, in view of the immediacy of their personal and social cost. The corresponding standard of 50 per million induced, viable chromosome anomalies and 2 per million dominant mutations entails a raw social cost of over $100 million. It is probably at the margin of ultimate detectability even by extrapolation from experimental assays; so that at least a billion-dollar argument needs to be put up by the defendant of any additive that gives a positive experimental score in such tests.

The costs of recessive mutations are much more difficult to estimate, being quite sensitive to the proportion of the *mutational* to the *segregational* load. At equilibrium, a 1 percent increase in the mutation rate will generate an estimated *economic* loss of about $1 billion per year (measured in the 1970 economy of the United States) but taking at least ten generations to approach full impact.

An extremely conservative estimate would then put the near-term annual tax connected to this standard at about $100 million for the recessives. These calculations give no weight to such costs (or savings) as may attend the gradual deterioration of intelligence and other complex functions as a consequence of cumulative genetic damage. Nor do they put a value on heartache.

These estimates are surely subject to an uncertainty of a factor of ten or so. They predicate the value of a human life as between $50,000 and $1 million per capita, depending on the age at which disability or death occurs and the level of custodial care entailed by it, as well as loss of economic productivity. They assign no value to early prenatal losses, though some would regard these as beneficial for the aim of zero population growth. This kind of cost-accounting is morally insufferable, but we must find some de facto standard of value in making hard decisions. If lives are valued at much more than a "million per bod," there is little evidence of this from the pragmatic behavior of the community or of most individuals in the choices they make in their daily lives.[9] However, these choices are made in a hindered market where the cost of safety is a side issue, more often obscured than intelligently ventilated.

On the other hand, a pesticide manufacturer would fire his director of public relations if he were to advertise that he calculated a life at less than his own annual income.

A health cost (from the "acceptable" standard) of $200 million per year is a grievous burden in absolute terms, but is immediately lost in an overall budget of over $100 billion. (Of this, $60 billion is direct health care; the indirect economic costs of disease, injury, and premature senescence are open-ended.) This is to say that a level of risk that approaches the intolerable, once we are well aware of it, may be impossible to verify by direct measurements of disease diffused throughout the population! In exceptional circumstances, an effect like the peculiar malformations induced by thalidomide comes to the surface, and then achieves a visibility and notoriety all out of proportion to other agents. *If the malformation induced by thalidomide were a mental retardation of 10 percent of the I.Q., instead of a highly characteristic and unusual deformation of the limbs, in an equal number of subjects, we would be unaware of it to this day.*

All this is to say that we must look to extrapolations from laboratory measures for the *only* reliable indication of mutagenicity in the human population!

Someday, it will be argued that the standard risk

should be elevated in a particular case, for example, were there to be a demonstrable net social benefit of, say, $1 billion per year from the use of an agent that elevated the mutation rate by 5 percent. The argument should not be rejected out of hand. For example, I believe that an acceleration of health research by $1 billion per year would improve the genetic and the overall health climate so as to more than outweigh the penalties of more mutations. If there were a harmonious redistribution of the resource benefit, we could foresee an advantageous tradeoff. The problem is to produce that harmony, to ensure that the people who bear the risk and eventually pay the price will also reap the benefits. Perhaps we will invent a tax on pesticides, earmarked for compensatory research. This makes sense only if we have exhausted alternative sources of income for such restorative purposes.

Should pesticides be particularly suspect as mutagens? Or would their pesticidal activity bear only an accidental relationship to genetic damage? One should answer this in relation to how much we know about the mechanism of action of the compound, and particularly the basis of its *specificity*. The fundamental cellular processes of all organisms are remarkably similar, the more so the closer we go to the genetic foundations. The DNA of the bacterium, the insect, the weed plant, the rodent pest, and of man has precisely the same architecture. For this reason, the most suspicious agents should be the disinfectants, the compounds that act directly on cells, and indeed most of these probably do act on the DNA of the microbe as the target. Other pesticides may also act on DNA, but owe their specificity to details of penetration, or of metabolism in the pest which will deliver the final toxic molecule to the cellular target. Finally, most pesticides probably act as enzyme inhibitors, but may have mutagenic effects as (1) incidental side effects of their own structure; (2) further metabolism in the mammalian system to genetically active products, or (3) as side effects of the inhibition of cellular enzymes. At the very least, pesticidal action does not disqualify a compound as a mutagen, and in some cases this may be very closely related to its intended mode of action. Only empirical studies, of the kind outlined in this monograph, can give a conclusive answer.

*These approaches are ingenious and already well enough calibrated to be a sensible basis of regulatory policy—at minimum the routine screening of every proposed new compound through several tests.*

However, this is only a start.

I do not believe that any routine arbitrary procedure will cover all of the potential hazards involved in extrapolating from laboratory data to man. Many new discoveries, some of fundamental importance to biological theory, will be made from unanticipated findings of harmful biological effects—just as the discovery of genetic damage by x-rays was a flash of creative genius when first surmised by H. J. Muller. In most cases it will be metabolic products of a pesticide that will cause trouble, not the initial compound, and it is imperative that we have a clear picture of their biochemical pathways in man. With such data, an investigator might, for example, realize that some biochemical deviants in the human population will be uniquely sensitive; or that the pesticide will interact with some drug or other environmental additive; or that some stage in fetal life may be uniquely sensitive. In any case, the system of evaluation must display all that is known about a new compound so as to assure the most creative thinking by pluralistic critics. This recommendation runs counter to existing policy which regards safety-evaluation data on a pesticide as the private property of the sponsor.

This purpose is probably unachievable, despite the most socially minded intentions, in the context of sponsor-directed research. I would advocate, instead, that the responsibility for safety-testing be in the hands of disinterested third parties, funded by fees taxed to the sponsors. This would also allow a fair allocation of costs to the coattail riders who will jump into competition now only after a pioneer has paid the initial costs of an evaluation out of his own pocket. It would also afford some leeway in selecting which tests are most likely to elicit damning news, a discretion that obviously cannot be vested in interested sponsors.

When chemical mutagenesis was a matter of speculation 30 years ago, many geneticists believed that its realization was indispensable to understanding the chemistry of the genetic material. History has revealed the opposite—other biophysical studies have given us the main clues about the structure of DNA, and we are just beginning to understand the complexity of the cellular processes that result in mutation in the light of that knowledge. We still do not know the local chemical change involved in radiation damage to DNA, while mutagenesis by base-substitution has been analyzed in so much detail that it is the textbook example of molecular pharmacology.[5] But for many other chemical mutagens there is much more to be learned, perhaps mostly in the metabolism and repair or misrepair of the DNA once it has suffered its primary damage.

This kind of insight already has great promise for more powerful assays of mutagenic potentiality in human cells, for example by direct measurement of the extent of DNA repair in cells subject to experimental or environmental insult, along the lines of Cleaver's studies on repair-deficient genotypes of man.[2]

Geneticists must not now overlook the other side of the coin—the enormous value of reliable measures to *decrease* the spontaneous mutation rate. There has been very little followup of the pioneering work on anti-mutagenic chemicals by Novick and Szilard two decades ago.[6]

Joshua Lederberg

References

1. Auerbach, Charlotte and Robson, J. M.
1944. Production of mutations by allyl isothiocyanate. *Nature 154*: 81.

2. Cleaver, J. E.
1969. Xeroderma Pigmentosum: A Human Disease in Which an Initial Stage of DNA Repair is Defective. *Proc. N.A.S., 63*: 428–435.

3. Some of the history of this incident is recounted by R. Egeberg, et al. 1970, Report of the Medical Advisory Group on Cyclamates, *J.A.M.A., 211*: 1358–1361. See also Mintz, Morton, Rise and Fall of Cyclamates, *Washington Post,* Oct. 26, 1969 (reprinted in the Congressional Record, Oct. 27, 1969, S–13256–8).

4. Federal Radiation Council.
May 13, 1960. Radiation Protection Standards Report No. 1. See also Federal Register, 4402, May 18, 1960.

5. Goldstein, Avram, Aronow, Lewis, and Kalman, S. M.
1968. Principles of Drug Action. Harper and Row, N.Y.

6. Novick, A., and Szilard, L.
1952. Antimutagens. *Nature 170*: 962–927.

7. Rusch, H. P., Bosch, Dorothy, and Boutwell, R. K.
1955. The influence of irritants on mitotic activity and tumor formation in mouse epidermis. *Acta Un. Int. Canc., 11*: 699–703.

8. Scientific Committee on the Effects of Atomic Radiation.
1966. Report to U.N. General Assembly, Official Records, 21st session. Suppl. 14 (A/6314).

9. Starr, C.
1969. Social benefit versus technological risk. *Science 165*: 1232–1238.

# The Mutagenicity of Pesticides
## Concepts and Evaluation

# 1 Introduction

This monograph is based on the Report of the Advisory Panel on Mutagenicity of Pesticides to the Secretary's Commission on Pesticides and Their Relationship to Environmental Health, HEW (GPO, December 1969). The advisory panel was cosponsored by the Environmental Mutagen Society. This is one of eight reports from subcommittees and advisory panels upon which the Secretary's Commission on Pesticides and Their Relationship to Environmental Health based its recommendations. Since each report was prepared by the membership of the subcommittee or advisory panel involved with the particular subject under review, this report by itself does not necessarily reflect the unanimous opinion of the commission's entire membership.

The Advisory Panel on Mutagenicity of Pesticides was constituted as follows: S. S. Epstein (Chairman), M. Legator (Executive Secretary), J. Crow, E. Freese, H. Malling, J. Neumeyer, W. Nichols, C. Wilkinson (Liaison), A. C. Kolbye (HEW staff), I. Gerring (HEW staff), and R. S. McCutcheon (HEW staff). Non-member contributors included S. Abrahamson, P. J. Bottino, R. Donnelly, L. A. Schaierer, A. H. Sparrow, and J. S. Wassom.

The final HEW Report on Mutagenicity of Pesticides

reflected the unanimous views and recommendations of the entire panel. It is, however, appropriate to identify contributors with major responsibility for initial preparation of various sections of the HEW report, and thus also of the corresponding sections of this monograph: S. Abrahamson, *Drosophila* (in chapter 3); J. Crow, Importance of Mutagenicity as a Potential Public Health Hazard (chapter 2); Population Monitoring (in chapter 3); S. S. Epstein, Dominant Lethal Test (in chapter 3); E. Freese and R. Donnelly, Bacterial Methods (in chapter 3); Structure Activity Relations of Pesticides (chapter 5); M. Legator, Host-Mediated Assay (in chapter 3); H. Malling and J. S. Wassom, Specific Locus Test (in chapter 3) and bibliography and literature review (in Appendix); J. Neumeyer, Usage Patterns of Pesticides (chapter 6) and tabulations (in Appendix); W. Nichols, Cytogenetic and Somatic Cell Genetics (in chapter 3); H. Sparrow, P. J. Bottino, and L. A. Schairer, Plants (in chapter 3).

Major modifications of the original HEW report, as embodied in this monograph, include the addition of a preface and rearrangement of bibliographies, amplification of the tabulations on pesticides, and addition of a cross index of pesticides.

## 2 Importance of Mutagenicity as a Potential Public Health Hazard

A particularly subtle danger from wide scale use of pesticides lies in the possibility that some may damage the hereditary material of man. If this is so, we may be unwittingly harming our descendants. Whether this is happening, and if so, what is the magnitude of the effect, are regrettably unknown. Surely, one of the greatest responsibilities of our generation is our temporary custody of the genetic heritage received from our ancestors. We must make every reasonable effort to ensure that this heritage is passed on to future generations undamaged. To do less, we believe, is grossly irresponsible.

The first evidence that environmental agents under human control might have some influence on the genetic constitution of future populations came with the discovery that high energy radiation causes mutations. The first convincing evidence of this came with the publication in 1927 of H. J. Muller's classic paper, "Artificial Transmutation of the Gene." Muller was quick to point out the potential health hazard associated with the indiscriminate use of radiation.

The discovery of nuclear energy brought a whole new dimension to the problem, and a greatly increased public awareness of genetic hazards. Out of this concern, originally confined largely to geneticists, radiologists, and radiation biologists but later including persons of a great diversity of special interests, have come safeguards to ensure that radiation exposure is kept to the lowest practicable minimum; whether these safeguards are adequate is now a subject of debate.

As soon as radiation-induced mutagenesis was discovered, there were strong reasons to suspect that many chemicals would have the same effect, but proof did not come until World War II, when mustard gas was shown to induce mutations in fruit flies. Since that time, efficient test systems have been developed, and a large number of chemicals of a great diversity of structure and activity have been shown to be mutagenic. The likelihood that some highly mutagenic chemicals may come into wide use, or indeed may already be in wide use, is great enough to be a cause for real concern. The fuel additive trimethyl phosphate is illustrative of a mutagenic chemical currently in extensive use.[1]

Pesticides are only a part of the many new chemical compounds that have entered our environment, but they are of particular concern because they are used so widely and in such enormous amounts. Furthermore, they are very potent biologically; if they were not, they would not be effective pesticides. The mechanisms by which most pesticides kill, inhibit growth, or sterilize the various animal and plant pests for which they are designed are thought to be unrelated to genetic mechanisms. Nevertheless, our ignorance of chemical mutagenesis will not allow us to assume that pesticides are safe without specific mutagenic tests.

What are mutations, and what effects will they have on the human population? In its broadest usage, the word *mutation* is used to designate any inherited change in the genetic material. The change may be a chemical transformation of an individual gene that causes it to have an altered function. Or it may involve a rearrangement, or a gain or loss, of parts of a chromosome. This kind of change is often visible by ordinary microscopy. We shall use the words *gene* or *point mutation* to designate changes of the individual gene, and speak of those changes which involve the larger chromosomal units as *chromosome aberrations*. In many experimental systems these are easily distinguished, but in human studies, classification of an individual defect as to whether it is due to a point mutation or a chromosome aberration is not always possible.

Mutations may occur anywhere in the body. Frequently, the result is the death of the particular cell in which the change occurs. Usually this causes only local and transient damage; for most individual cells are quite dispensable. But if the change affects the genetic functioning of the cell while still permitting it to divide, this change may be transmitted to descendant cells and so will cause less localized damage. The effect may be cancerous or it may be teratogenic, particularly if the change takes place during embryonic development. We are especially interested, of course, in those changes that occur in the germ cells—cells that are the progenitors of future generations. A mutation or chromosome change, transmitted via the sperm or egg to the next generation, can affect every cell in the body of the descendant individual, with consequences that may be disastrous.

What kinds of effects on the human being do mutations produce? Perhaps the most important fact to emphasize is that there is no single effect. Since every part of the body and every metabolic process are influenced by genes to a greater or lesser extent, it comes as no surprise that the range of effects produced by gene alterations includes every kind of structure and process.

At one extreme are consequences—so-called lethal effects—so severe that the individual cannot survive. If the death occurs very early in embryonic development, it may never be detected. If the death is at a later stage, it may lead to a miscarriage. An appreciable fraction, very roughly one-fourth, of spontaneous abortions shows a detectable chromosome aberration, and there is no way at present to know how many of the remainder are caused by gene mutations or by chromosome aberrations too small to detect by the microscope. If the embryo survives until birth, there may be physical abnormalities. There are hundreds of known inherited diseases, and probably many more that are unknown, all of which owe their ultimate cause to mutations. These are individually rare, but collectively they account for a substantial fraction of human misery. And, perhaps most tragic of all, genetic factors play a role in the causation of mental deficiency and mental disease.

At the other extreme are consequences so mild that they finally become imperceptible. Between these extremes are the whole gamut of genetic defects, from minor to severe. So, it is evident that the effect of an increased rate of mutation and chromosome aberration is not something new, but reflects, rather an increased frequency of diseases, abnormalities, weaknesses, and assorted human frailties that are already occurring.

Many mutations produce effects similar to those produced by other nongenetic causes. And we must remember that spontaneous mutations are happening all the time. For these reasons, the impact of environmental mutagens is statistical rather than unique. This problem is further complicated by the time-distribution of mutational effects. Some mutant genes are dominant, in which case the abnormality or disease will appear in the very next generation after the mutation occurs. On the other hand, the gene may be recessive; that is to say, it may require the abnormal genes in both homologous chromosomes (one derived from the male parent and the other from the female parent) to produce the effect. In this case, the disease or abnormality may be delayed for many generations, until some unlucky child inherits a mutant gene simultaneously from each of his parents. The net result of all this is that, although the first generation probably has a larger effect than any particular subsequent generation, the overall effect is spread over many generations. What happens in the first generation is only a fraction of the total impact of the mutation process.

That the great majority of mutations should be harmful to a greater or lesser extent (or if not harmful, at best neutral) is both a deduction from the principle of natural selection and an empirical fact well established in experimental organisms. In the human past, natural selection has ruthlessly eliminated those individuals whose mutant genes caused them to be abnormal, diseased, or even only slightly weakened. As a result, there has been an approximate equilibrium between the introduction of new mutant genes into the population by mutation and the elimination of old genes by natural selection. But with our present high standards of living and health care, many mutants that, in the past, would have caused death or reduced fertility now persist. So the equilibrium is out of balance, and new mutants are being added to the population faster than they are being eliminated. This fact, coupled with the near eradication of many infectious diseases, means that now and in the future our medical problems will be increasingly of genetic origin.

A mutation, once it has occurred, is transmitted from parents to succeeding generations. If the gene causes a lethal or sterilizing effect, it will persist for only one generation and affect only one person. On the other hand, if it causes only a slight impairment, it may be transmitted from generation to generation and thereby affect many people. There is, therefore, generally an inverse relation between the severity of the gene effect and the number of persons that will be exposed to this effect. If it were not for this, we could dismiss the effects of mild mutants as relatively unimportant. But in any overall consideration, we must consider many persons, mildly affected, as being of comparable importance to one individual severely affected. Experiments on fruit flies show that mildly deleterious mutations occur with much greater frequency than more severe mutations—at least ten times as frequently. It is likely that, although an increased mutation rate would cause a corresponding increase in severe abnormalities and genetic diseases, the major statistical impact of a mutational increase on the human population would be to add to the burden of mild mutational effects. This would make the population weaker, more prone to disease, and more likely to succumb to an effect that otherwise would be resisted.

All these implications mean that it is not possible to predict in detail the kinds of effects that would occur following an increased rate of mutations, or their distribution in time. Nor can we be at all accurate in any quantitative assessment of the total harmful impact of mutation on the population, in comparison with other hazards. So, in weighing benefits against risks of possibly mutagenic pesticides, we have only a vague idea of the nature and magnitude of the risk. We must remember, however, that genetic damage is irreversible by any process known to us now. The risk to future generations, though difficult to assess in precise terms, is nevertheless very real. The prevention of any unnecessary mutational damage in the future is one of our most important and critical responsibilities today.

Despite the extensive use of pesticides, our information on their possible mutagenicity is grossly inadequate. Several have been tested in various test systems, but we believe that none has had the kind of systematic testing that would be regarded as minimally acceptable. Such testing, we believe, is entirely practical and feasible. There are numerous widely used test systems that are precise, efficient, and relatively inexpensive. However, these depend mainly on microbial, insect, or plant systems; there is a question as to their relevance to man. For definitive testing, it is essential to use systems that have a high degree of presumptive human relevance. We believe that satisfactory systems now exist that are practical, precise, and relevant. In this report, we recommend that a combination of these systems be routinely applied to all pesticides.

Reference

1. Epstein, S. S., Bass, W., Arnold, E., and Bishop, Y. 1970. The mutagenicity of trimethyl phosphate in mice. *Science 168*: 584–586.

# 3 Methods for Mutagenicity Testing

## Introduction

Various methods are now available for mutagenicity testing. From the criterion of presumptive human relevance, they have been categorized as ancillary submammalian systems and definitive mammalian systems. The human relevance of data obtained from ancillary test systems is uncertain, in view of factors such as cell uptake, metabolism, detoxification, dosage, and method of administration. Mammalian systems entail fewer of these limitations.

Since no single method can detect all possible types of mutations, a combination of methods must be used. A positive result in *any* mammalian system represents evidence of a potential mutagenic hazard. The danger inherent in the use of restricted and inappropriate test systems is apparent from recent NIH contract-supported studies in which mutagenic activity of pesticides was tested in microbial systems. In these studies, the microbial systems could have only detected point mutations, whereas structural considerations indicated that the pesticides tested could only induce inactivating DNA alterations resulting in chromosome breaks and aberrations. Additionally, some of the pesticides tested required microsomal enzymatic activation, which could only occur in vivo mammalian test systems.

In addition to these test procedures, the monitoring of human populations may reveal mutagenic effects of pesticides, or any other environmental agents, that have escaped detection.

## Ancillary Methods

### Bacterial Methods

A variety of relatively simple and inexpensive tests are available for demonstrating point mutations. Reverse and forward mutations may be scored in bacteria using nutritional, resistance, or fermentative markers. The best systems, however, are those based on the use of biochemically and genetically characterized bacterial strains. The action of specific mutagens may then be better understood by analysis of changes induced at the molecular level.

The reverse mutation system of *Salmonella* is an excellent example of a genetically defined bacterial mutant.[3] Many histidine-requiring mutants exist which revert by single base pair changes, i.e., transitions or base pair insertions or deletions. By selection of proper strains, most possible mutations may be detected in a simple plate test. The histidine-requiring mutation is challenged on a synthetic agar plate, and effects of mutagens may be recognized by the presence of histidine-independent mutant colonies. Another example is the use of genetically characterized *E. coli* K–12 strains.[1] These are being currently used to analyze the mutagenic action of a variety of chemicals.[2]

In addition to *Salmonella* and *E. coli* systems, a variety of other bacterial systems are being used to correlate the mode of action of a particular agent with a specific genetic response. The genetic fine mapping and biochemical analyses among several bacterial sites far exceeds that known for higher organisms; added to this is the obvious fact that bacteria—because of this ease of handling of huge populations, rapid mode of replication, and multidetectable mutation sites—lend themselves ideally for large scale evaluation of mutagens.

References
1. Mukai, E., Hawryluk, I., and Shapiro, R. 1970. The mutagenic specificity of sodium bisulfite. *Biochem. Biophys. Res. Comm.*

2. Whitfield, H. J., Martin, R., and Ames, D. J. 1966. Classification of amino-transferance (C-gene) mutants in the histidine operon. *Mol. Biol. 21*: 335–355.

3. Yanofsky, C., Cox, E. C., and Horn, V. 1966. The unusual mutagenic specificity of an *E. coli* mutation gene. *Proc. Nat. Acad. Sci. 55*: 274–281.

## Neurospora

*Neurospora crassa* is a haploid organism with seven chromosomes and a normal meiotic cycle. However, by using a balanced heterokaryon between biochemically marked strains, the diploid phase of higher organisms can be mimicked. Chromosome deletions as well as point mutations can thus be detected. Forward mutations can be recovered in the ad–3 region of chromosome 1[1] without applying selective techniques. Either growing cultures or spores (*conidia*) can be exposed to chemicals under test. After the treatment, the *conidia* are inoculated into 10-liter Florentine flasks and incubated for seven days. Each flask can contain $10^6$ colonies, which are screened for presence of purple mutants. The frequency of the different fractions of the *conidia* population from the heterokaryon can be determined by plating on different substrates.

Refined genetic analysis can be carried out on the mutants. The frequency and the size of the chromosome deletion can be determined,[5] and the genetic alterations of the point mutations can be identified at the molecular level.[3, 4] From the plate counts, it is possible to distinguish between nuclear inactivation and cytoplasmic inactivation.

Mutational frequencies induced by 500 r can be easily and practically detected. *Neurospora* is obviously metabolically different from mammals; therefore, tests for mutagenicity should include mammalian metabolites of the pesticide and the use of *Neurospora* in the host-mediated assay.[2]

References

1. de Serres, F. J., and Osterbind, R. S.
1962. Estimation of the relative frequencies of X-ray-induced viable and recessive lethal mutations in the ad–3 region of *Neurospora crassa*. *Genetics 47*: 793–796.

2. Gabridge, M., and Legator, M. S.
1969. A host-mediated microbial assay for detection of mutagenic compounds. *Proc. Soc. Exp. Biol. Med. 130*: 831–834.

3. Malling, H. V., and de Serres, F. J.
1965. Identification of the genetic alterations in nitrous acid-induced ad–3 mutants of *Neurospora crassa. Mutation Res. 2*: 320–327.

4. Malling, H. V., and de Serres, F. J.
1967. Relation between complementation patterns and genetic alterations in nitrous acid-induced ad–3B mutants of *Neurospora crassa. Mutation Res. 4*: 425–440.

5. Webber, B. B., and de Serres, F. J.
1965. Induction kinetics and genetic analysis of X-ray induced mutations in the ad–3 region of *Neurospora crassa. Proc. Natl. Acad. Sci. U.S. 53*: 430–437.



## Phage and Transformation

**Phage[1]** Bacteriophage $T_4$ is probably the best available system. Forward mutations to the "r" phenotype are of low sensitivity; reverse mutations of "rII"–type mutants are of high sensitivity. Chemicals inducing point mutations, which alter DNA either chemically—treatment of free phages—or during its duplication inside the bacteria, can be detected. Sensitive assays based on reverse mutations respond only to those agents which induce the required specific base pair change, e.g., $\frac{G}{C} \rightarrow \frac{A}{T}$. In order to detect all types of base pair changes, a set of about six rII mutant strains having the required base pair changes should be tested. Agents which induce only inactivating DNA alterations rarely induce point mutations. They do, however, inactivate phage, but only more detailed genetic tests can verify that the inactivation is not caused by an alteration of phage protein.

**Transformation**[1] The ideal system is that of linked mutation induction, which, at present, is limited to the induction of fluorescent mutants in the tryptophan operon. Forward mutations to fluorescence are of medium sensitivity. Reverse mutations to indole independence are of high sensitivity. In these systems, inactivating DNA alterations can be measured and quantitatively compared to mutagenic DNA alterations. It has been shown that radical-producing agents, known to induce both chromosome breaks and large chromosome mutations, inactivate transforming DNA but do not induce point mutations. Thus, in most bacterial or phage systems, these agents would not induce mutations, and might be erroneously labelled nonmutagenic. Only agents which act directly on resting DNA can be easily assayed. For agents, like base analogs, which induce mutations in duplicating DNA, such measurements are difficult.

Reference

1. Freese, E., and Freese, E. B.
1966. Mutagenic and inactivating DNA alterations. *Radiation Res. 6* (*Suppl.*): 97–140.

## Plants

An extensive literature exists on the response of higher plants to chemical mutagens. Many techniques developed for studies with physical mutagens should be equally applicable to known or suspected chemical mutagens.[26, 42, 56, 90]

The literature indicates that many well-known mutagenic and/or chromosome-breaking chemicals are as effective in plant as in animal test systems. Several plant systems can detect effects of such substances in the gaseous state.[81] Some, if not all, plants are highly susceptible to chemical mutagens. Thus, appropriate plant systems should be included in the battery of tests for mutagenicity of various chemicals. Further, since plant chromosomes are structurally more akin to mammalian chromosomes than are those of viruses, bacteria, and other prokaryotic organisms, responses of plant chromosomes to chemical mutagens should provide valuable information with respect to their possible mutagenicity in mammals. Also, the factors determining the inherent radiosensitivity of plant cells are now fairly well understood.[87]. This knowledge may offer useful guidance for work with chemical mutagens.

Plant systems utilize many species and a considerable

variety of possible procedures at various stages of development (Table 3.1). It is not feasible to select a specific test as the best in all possible cases. Circumstances and objectives of the experiment would determine which test and which species should be recommended.

A brief outline follows of the procedures and specific plants considered the most promising for chemical mutagen studies.

**Mutation Induction by Seed Treatment** Barley (*Hordeum*) has been used extensively in the study of induced mutations and can be recommended because of the extensive knowledge of its genetics, including numerous *and* distinct chlorophyll-deficient mutations,[66] and because of the small number ($2n = 14$) of its relatively large chromosomes. The seed is very easy to store, treat, and handle, and the seedlings are small and easy to grow. Measurable responses to mutagen treatment include (1) chromosome aberrations in somatic or meiotic cells, (2) chlorophyll-deficient mutations, (3) pollen abortion, (4) mutation spectrum of $M_2$ seedlings, (5) seedling growth reduction, (6) survival, and (7) spike fertility.

**Specific-Locus Method (*Waxy* Locus) in Pollen** The *waxy* locus in maize, barley, and rice determines the type of starch which is synthesized in the triploid endosperm and in the haploid pollen grain. In the case of pollen, the phenotype is determined by its own genotype and not by the genotype of the parent plant. The dominant *Wx* pollen grains stain blue with an iodine-potassium iodide stain while the recessive *wx* pollen grains stain a reddish-brown color. Since the wild type is *Wx*, the frequency of induction of *wx* can be assayed in millions of pollen grains relatively easily and quickly. Furthermore, the phenotype appears in the treated generation, and thus it does not require the time necessary to obtain an $M_2$ generation. This is a very simple and rapid technique for detecting even very low frequencies of induced mutations in higher plants, and it should be amenable to automated scoring procedures.

**Root-tip Method for Chromosome Aberrations** Root tips of certain species, e.g., *Allium cepa*,[12] *Tradescantia*,[9] and *Vicia faba*[9] provide excellent material and have been used extensively for work with chemical mutagens.[56] These species are easy to obtain and to grow, and their root tips, readily treated with aqueous solutions, provide large cell populations. Treatment periods are short (minutes to hours) and fixations and squashes are usually made within 48 hours. Although the slides are currently screened by eye, computer-assisted analysis could be used.

**Somatic Mutation Methods** *Tradescantia* plants, heterozygous for flower color, also provide a useful test system which can be exposed as young flower buds to various mutagens in either a gaseous or an aqueous state. The plants are easy to grow under a wide range of environmental conditions, bloom throughout the year, thus providing a continuous supply of material, and have radiosensitivity similar to that of mammalian cells. Both petals and stamen hairs can be simply scored for a 5 to 15-day period using only a dissecting microscope.[38, 64, 83]

Various other genera and species should prove useful in chemical mutagen studies (Table 3.1).

A partial list of chemical mutagens and/or pesticides known to be effective in higher plants is given in Tables 3.2 and 3.3. Several other pesticides are also known to be mutagenic in plants, e.g. Amitrole,[96] Isocil,[96] Lindane,[78] and Dichlorvos.[78] An indication of relative mutation rates produced by gamma rays and various chemical mutagens in several test systems is given in Table 3.4.

**Table 3.1** Summary of Experimental Procedures for Detecting Various Effects of Chemical Mutagens Using Certain Higher-Plant Test Systems.

| Stage Treated | Analyses Used for Detecting Various Effects | | | | | | | | Species Used and Literature Reference |
|---|---|---|---|---|---|---|---|---|---|
| | Somatic Cells | | Germinal Cells | | Microspores or Pollen Tubes | Pollen | | | |
| | | | | | | Mutation | | Chr. Ab.† | |
| | Mut. | Chr. Ab. | Mut. | Chr. Ab. | Chr. Ab. | Phenotype** | Lethal | Translocation | |
| Seed or seedling | + | + | P‡ | + | +¶ | + | – | + | *Allium*, 54, 78 *Arabidopsis*, 34, 60, 73 *Avena*, 35 *Hordeum*, 17, 18, 50, 52, 66, 93, 96 *Oryza*, 71, 97 *Triticum*, 61 *Zea*, 80 |
| Flowering | | | | | | | | | |
| Somatic cells* | + | + | – | – | – | – | – | – | *Antirrhinum*, 84 *Cosmos*, 27 *Dianthus*, 10 *Gladiolus*, 83 *Haemanthus*, 63 *Lilium*, 84 *Petunia*, 84 *Tradescantia*, 13, 36, 38, 39, 61, 64, 83 *Tulipa*, 83 |
| Meiosis | – | – | P | + | + | + | + | – | *Hordeum*, 19 *Tradescantia*, 29, 77 *Trillium*, 86 *Tulipa*, 65 *Vicia*, 59 *Zea*, 20 |
| Microspore | – | – | – | – | + | – | – | +¶ | *Campelia*, 76 *Tradescantia*, 3, 21, 22, 77, 85, 90 |
| Pollen | – | – | P | – | + | – | – | – | *Hordeum*, 57 *Lilium*, 4 *Petunia*, 4 *Tradescantia*, 4, 81 *Zea*, 4 |

*Mostly petals and stamen hairs; †Translocations produce pollen abortion; ‡P = progeny testing; **Especially waxy locus in barley, maize, and rice; ¶ Micronuclei can be counted as a fast screening method.

Table 3.2 Partial List of Chemicals Known to Produce Mutations or Chromosome Aberrations in Higher Plants with Literature Citations.

| Chemical | Mutation | Chromosome Aberration |
|---|---|---|
| Acridine (and derivatives) | | 9 |
| **Alkaloids** | | |
| colchine | 74 | 70 |
| morphine | | 70 |
| scopalamine | | 70 |
| **Amines and Related Compounds** | | |
| acetylethyleneimine | | 69 |
| 2-chlorotriethyl-amine | 52 | |
| ethyleneimine | 78 | |
| hydrazine | 39 | |
| hydroxyurea | | 45 |
| maleic hydrazide | 25 | 12, 25 |
| N-methylphenyl-nitrosamine | | 44 |
| nitrosamines | 92 | |
| triethylene melamine | | 6 |
| **Antibiotics and Related Compounds** | | |

| Chemical | Mutation | Chromosome Aberration |
|---|---|---|
| aminopterin | 60 | 43 |
| streptonigrin | | 47 |
| nebularine (9-β-D-ribofuranosylpurine) | 95 | |
| bromine | | 7 |
| Ceepryn | | 82 |
| 2,2-dichlorovinyl dimethyl phosphate (Vapona) | | 78 |
| diethyl sulfate | 30, 52 | |
| **Epoxides** | | |
| diepoxybutane | 17 | |
| ethylene oxide | 30, 88, 95 | 81 |
| glycidol | 15 | |
| **Food additives** | | |
| butylated hydroxy toluene | | 78 |
| butylated hydroxyanisole | | 78 |
| coumarin | | 8 |
| sucaryl | | 78 |
| hexachlorocyclo-hexane | | 53 |

| Chemical | Mutation | Chromosome Aberration |
|---|---|---|
| isopropylphenyl carbamate | | 14 |
| **Mercury compounds** | | |
| ethyl mercuric phosphate | | 75 |
| methyl mercuric hydroxide | | 72 |
| phenyl mercuric hydroxide | 58 | 58, 72 |
| **Mustards** | | |
| sulfur mustard | | 11 |
| nitrogen mustard | 33, 55 | 68, 115 |
| **Nucleosides** | | |
| adenine arabinoside (arabinosyl-adenine) | | 48 |
| adenine xyloside (xylosyl-adenine) | | 48 |
| 5-bromodeoxy-cytidine | 34 | |
| 5-bromodeoxy-uridine (BUdR) | 34 | |
| deoxyadenosine | | 43 |
| cytosine arabinoside | | 43 |

| Chemical | Mutation | Chromosome Aberration |
|---|---|---|
| 5-fluorodeoxyuridine (FUdR) | | 43, 45, 91 |
| 1-methyl-3-nitro-1-nitrosoguanidine | | 40 |
| N-methyl-N′-nitrosoguanidine | 79 | |
| **Pesticides (see Table 3.3)** | | |
| $\beta$-propiolactone | | 82, 89 |
| **Purine** | | |
| 2-aminopurine | 28, 49 | |
| caffeine (1,3,7-trimethylxanthine) | | 46 |
| 8-ethoxycaffeine | | 41, 47 |
| 1,3,7,9-tetramethyl-uric acid | | 47 |
| **Sulfonic acids and derivatives** | | |
| methane sulfonates | | |
| bromoethyl | 32 | |
| n-butyl | 52, 67 | |
| chloroethyl | 32 | |
| ethyl | 1, 2, 4, 73 | |
| $\beta$-hydroxyethyl | 24 | |

| Chemical | Mutation | Chromosome Aberration |
|---|---|---|
| β-methoxyethyl | 24 | |
| methyl | 52, 62 | |
| isopropyl | 52, 67 | |
| n-propyl | 16, 52 | |
| β-methane sulfonyl oxybutane (Myleran) | 26, 28, 95 | 95 |
| diethyl 1,3-propane-disulfonate | 24 | |
| o-sulfobenzoicimide (Saccharin) | | 78 |
| **Urethanes** | | |
| ethyl | | 70 |
| N-nitroso-N-methyl | | 44 |

**Table 3.3** List of Various Pesticides (1000 ppm, 12 hours) Known to Produce Mutations in Barley and Relative Efficiency of Each to Control and to 5500 R of X-rays (Wuu and Grant, 96)

| Treatment | Relative efficiency |
|---|---|
| Lorox | 30 |
| Simazine | 24 |
| ENT-50612 | 14 |
| Atrazine | 10 |
| Monuron | 10 |
| Embutox E | 9 |
| Sevin | 9 |
| Banvel D | 7 |
| Botran | 7 |
| Phosphamidon | 7 |
| Alanap-3 | 4 |
| Metepa | 4 |
| Endrin | 3 |
| X-rays | 32 |
| Control | 1 |

**Table 3.4** Maximum Percent Mutations Reported for a Series of Mutagens Tested on Several Organisms (52).

| Agents | Mutations at Several Loci | | Mutations at Specific Loci | |
|---|---|---|---|---|
| | Barley (*Hordeum*) Chlorophyll Mutants mutated spikes* | Drosophila Sex-linked rec. lethals† | *Neurospora ad* reversions‡ | *Schizo. pombe arg* reversions** |
| Gamma rays | 17 | | | |
| Diethyl sulfate (DES) | 43 | | 18 | 0.1200 |
| Methyl methanesulonate (MMS) | 33 | 11.6 | | 0.0220 |
| Ethyl methanesulfonate (EMS) | 57 | 39.0 | 17 | 0.0910 |
| Chloroethyl methanesulfonate | | | 51 | 0.0220 |
| n-Propyl methanesulfonate (nPMS) | 26 | | | |
| Isopropyl methanesulfonate (isoPMS) | 20 | | | |
| n-Butyl methanesulfonate (nBMS) | 28 | 0.8 | | |
| Ethyl ethanesulfonate (EES) | 25 | | | |
| Nitrogen mustards | | | | |
| 2-chlorotriethylamine | 15 | | | 0.0009 |
| chlorodimethylamine | | | 1.7 | |
| Ethyleneimine | 28 | | 16 | |
| Diepoxybutane | | | 85 | 0.0160 |
| Glycidol | 22 | | 34 | 0.0240 |
| Ethylene oxide | 13 | | 17 | |

*After compilation by Nilan et al. (66, 67); †Data from Fahmy and Fahmy (23); ‡After compilation by Westergaard (94); **Data from Heslot (31).

References


1. Amano, E.
1967. Comparison of ethyl methanesulfonate- and radiation-induced *waxy* mutants in maize. *Mutation Res. 5*: 41–46

2. ———, and Smith, H. H.
1965. Mutations induced by ethyl methanesulfonate in maize. *Mutation Res. 2*: 344–351.

3. Beatty, A. V., and Beatty, J. W.
1967. Radiation repair of chromosome breaks as effected by constituents of nucleic acids. *Radiation Botany 7*: 29–34.

4. Brewbaker, J. L., and Emery, G. C.
1961. Pollen radiobotany. *Radiation Botany 1*: 101–154.

5. Briggs, R. W., Amano, E., and Smith, H. H.
1965. Genetic recombination with ethyl-methane-sulphonate-induced *waxy* mutants in maize. *Nature 207*: 890–891.

6. Buiatti, M., and Ronchi, V. N.
1963. Chromosome breakage by triethylemelamine (TEM) in *Vicia faba* in relation to the mitotic cycle, *Caryologia 16*: 397–403.

7. Chury, J., and Slouka V.
1949. Effects of bromine on mitosis in root tips of *Allium cepa*. *Nature 163*: 27–28.

8. D'Amato, F., and D'Amato-Avanzi, M. G.
1954. The chromosome-breaking effect of coumarin derivatives in the *Allium* test. *Caryologia 6*: 134–150.

9. ———, and Avanzi, S.
1954. Quarto contríbuto alla conoscenza dell' attività mutagena dei derivati dell'acridina. *Caryologia 6*: 77–89.

10. ———, Moschini, E., and Pacini, L.
1964. Mutazioni somatiche nel garofano indotte dalla radiazione gamma. *Caryologia 17*: 93–101.

11. Darlington, C. D., and Koller, P.
1947. The chemical breakage of chromosomes. *Heredity 1*: 187–221.

12. ———, and McLeish, J.
1951. Action of maleic hydrazide on the cell. *Nature 167*: 407–408.

13. Davies, D. R.
1963. Radiation-induced chromosome aberrations and loss of reproductive integrity in *Tradescantia*. *Radiation Res., 20*: 726–740.

14. Doxey, D.
1949. The effect of isopropylphenyl carbamate on mitosis in rye and onion cells. *Ann. Botany* (London) *13*: 329–335.

15. Ehrenberg, L.
1960. Chemical mutagenesis: Biochemical and chemical points of view on mechanism of action. In *Chemische Mutagenese. Erwin-Baur-Gedächtnis-vorlesungen II*. Berlin: Akademie-Verlag, pp. 124–136.

16. ———.
1960. Induced mutation in plants: mechanisms and principles. *Genet. Agraria 12*: 364–389.

17. ———, and Gustafsson, Å.
1957. On the mutagenic action of ethylene oxide and diepoxybutane in barley. *Hereditas 43*: 595–602.

18. ———, Gustafsson, Å., and Lundqvist, U.
1961. Viable mutants induced in barley by ionizing radiations and chemical mutagens. *Hereditas 47*: 243–282.

19. Eriksson, G.
1962. Radiation-induced reversions of a *waxy* allele in barley. *Radiation Botany 2*: 35–39.

20. ———, and Tavrin, E.
1965. Variations in radiosensitivity during meiosis of pollen mother cells in maize. *Hereditas 54*: 156–169.

21. Evans, H. J.
1962. Chromosome aberrations induced by ionizing radiations. *Intern. Rev. Cytol. 13*: 221–321

22. ———.
1968. Repair and recovery at chromosome and cellular levels: Similarities and differences. *Brookhaven Symp. Biol. 20*: 111–133.

23. Fahmy, O. G., and Fahmy, M. J.
1957. Mutagenic response to the alkylmethanesulphonates during spermatogenesis in *Drosophila melanogaster. Nature 180*: 31–34.

24. Gichner, T., Ehrenberg, L., and Wachtmeister, C. A.
1968. The mutagenic activity of β-hydroxyethyl methanesulfonate, β-methoxyethyl methanesulfonate, and diethyl 1,3-propanedisulfonate. *Hereditas 59*: 253–262.

25. Grant, W. F., and Harvey, P. M.
1960. Cytogenic effects of maleic hydrazide treatment of tomato seed. *Can. J. Genet. Cytol. 2*: 162–174.

26. Gröber, K., Scholz, F., and Zacharias, M. (eds.).
1967. *Induzierte Mutationen und ihre Nutzung. Erwin-Baur-Gedächtnisvorlesungen IV*. Berlin: Akademie-Verlag, 463 pp.

27. Gupta, M. N., and Samata, Y.
1963. Somatic mutation studies on *Cosmos bipinnatus* with particular reference to induction of flower color changes. *Gamma Field Symposia 2*: 69–78.

28. Gustafsson, Å.
1960. Chemical mutagenesis in higher plants. In *Chemische Mutagenese. Erwin-Baur- Gedächtnisvorlesungen I.* Berlin: Akademie-Verlag. pp. 14–29.

29. Haque, A.
1953. The irradiation of meiosis in *Tradescantia*. *Heredity 6*: (Suppl.) 35–40.

30. Heslot, H.
1960. Induction de mutations chez les plantes cultivées. Recherches effectuées par quelques agronomes français. *Genet. Agraria 13*: 79–112.

31. ———
1961. Étude quantitative de revérsions biochimiques.

induites chez la levure *Schizosaccharomyces pombe* par des radiations et des substances radiomimétiques. In *Strahleninduzierte Mutagenese. Erwin-Baur-Gedächtnisvorlesungen II.* Berlin: Akademie-Verlag, pp. 193–228.

32. ———, Ferrary, R., Levy, R., and Monard, C. 1959. Recherches sur les substances mutagènes: (halogéno-2-éthyle) amines, dérivés oxygénés du sulfure de bis (chloro-2-éthyle), esters sulfoniques et sulfuriques. *Compt. Rend. Acad. Sci.* (Paris) *248*: 729–732.

33. ———. 1961. Induction de mutations chez l'orge: Efficacité relative des rayons gamma, du sulfate d'éthyle, du méthane sulfonate d'éthyle, et de quelques autres substances. In *Effects of Ionizing Radiations on Seeds.* Vienna: International Atomic Energy Agency, pp. 243–249.

34. Hirono, Y., and Smith, H. H. 1969. Mutations induced in *Arabidopsis* by DNA nucleoside analogs. *Genetics 61*: 191–199.

35. Ichikawa, S., and Ikushima, T. 1967. A developmental study of diploid oats by means of radiation-induced somatic mutations. *Radiation Botany 7*: 205–215.

36. ———, and Sparrow, A. H. 1967. Radiation-induced loss of reproductive integrity in the stamen hairs of a polyploid series of *Tradescantia* species. *Radiation Botany 7*: 429–441.

37. ———. 1968. The use of induced somatic mutations to study cell division rates in irradiated stamen hairs of *Tradescantia virginiana* L. *Japan. J. Genetics 43*: 57–63.

38. ———, and Thompson, K. H. 1969. Morphologically abnormal cells, somatic mutations and loss of reproductive integrity in irradiated *Tradescantia* stamen hairs. *Radiation Botany 9*: 195–211.

39. Jain, H. K., Raut, R. N., and Khamankar, Y. G. 1968. Base specific chemicals and mutation analysis in *Lycopersicon. Heredity 23*: 247–256.

40. Kaul, B. L. 1969. The effect of some treatment conditions on the radiomimetic activity of 1-methyl-3-nitro-1-nitrosoguanidine in plants. *Mutation Res. 7*: 43–49.

41. Kihlman, B. A. 1955. Oxygen and the production of chromosome aberrations by chemicals and x-rays. *Hereditas 41*: 384–404.

42. ———. 1966. *Actions of Chemicals on Dividing Cells.* Englewood Cliffs, N.J.: Prentice-Hall, 260 pp.

43. ———. 1966. Deoxyribonucleotide synthesis and chromosome breakage. In C. D. Darlington and K. R. Lewis (eds.), *Chromosomes Today*, Edinburgh: Oliver and

Boyd, Vol. 1, pp. 108–117.

44. ———, and Eriksson, T.
1962. The distribution between cell nuclei of isolocus breaks and chromatid interchanges induced by radiomimetic chemicals in *Vicia faba*. *Hereditas 48*: 520–529.

45. ———, and Odmark, G.
1966. Effects of hydroxyurea on chromosomes, cell division and nucleic acid synthesis in *Vicia faba*. *Hereditas 55*: 386–397.

46. ———, and Levan, A.
1949. The cytological effect of caffeine. *Hereditas 35*: 109–111.

47. ———, and Odmark, G.
1965. Deoxyribonucleic acid synthesis and the production of chromosomal aberrations by streptonigrin, 8-ethoxycaffeine, and 1,3,7,9-tetramethyluric acid. *Mutation Res. 2*: 494–505.

48. ———.
1966. Effects of adenine nucleosides on chromosomes, cell division, and nucleic acid synthesis in *Vicia faba*. *Hereditas 56*: 71–82.

49. Kleinhofs, A., Gorz, H. J., and Haskins, F. A.
1968. Mutation induction in *Melilotus alba annua* by chemical mutagens. *Crop Sci. 8*: 631–632.

50. Konzak, C. F., Nilan, R. A., Froese-Gertzen, E. G., and Foster, R. J.
1965. Factors affecting the biological action of mutagens. In *Induction of Mutations and the Mutation Process*. Prague: Czechoslovak Academy of Sciences, pp. 123–132.

51. ———, and Ramirez, I. A.
1966. Physical and chemical mutagens in wheat breeding. *Hereditas Suppl. 2*: 65–84.

52. Konzak, C. F., Nilan, R. A., Wagner, J., and Foster, R. J.
1965. Efficient chemical mutagenesis. *Radiation Botany 5* (Suppl.): 49–70.

53. Kostoff, D.
1949. Induction of cytogenetic changes and atypical growth by hexachlorocyclohexane. *Science 109*: 467–468.

54. Levan, A.
1951. Chemically induced chromosome reactions in *Allium cepa* and *Vicia faba*. *Cold Spring Harbor Symp. Quant. Biol. 16*: 233–243.

55. Loveless, A.
1951. Qualitative aspects of the chemistry and biology of radiomimetic (mutagenic) substances. *Nature 167*: 338–342.

56. ———.
1966. *Genetic and Allied Effects of Alkylating Agents.* London: Butterworths, 270 pp.

57. Lundqvist, U.
1964. Induction of mutations in barley pollen by ultraviolet and x-rays. In *Barley Genetics I. Proc. 1st*

*Intern. Barley Genetics Symp.* Wageningen, pp. 92–95.

58. MacFarlane, E. W. E.
1950. Somatic mutations caused by organic mercurials in flowering plants. *Genetics 35*: 122–123.

59. Marshak, A.
1939. A comparison of the sensitivity of mitotic and meiotic chromosomes of *Vicia faba* and its bearing on theories of crossing-over. *Proc. Natl. Acad. Sci. U.S. 25*: 510–516.

60. McKelvie, A. D.
1963. Studies in the induction of mutations in *Arabidopsis thaliana* (L.) Heynh. *Radiation Botany 3*: 105–123.

61. Mericle, L. W., and Mericle, R. P.
1965. Biological discrimination of differences in natural background radiation level. *Radiation Botany 5*: 475–492.

62. Minocha, J. L., and Arnason, T. J.
1962. Mutagenic effectiveness of ethyl methanesulfonate and methyl methanesulfonate in barley. *Nature 196*: 499.

63. Molè-Bajer, J.
1965. Telophase segregation of chromosomes and amitosis. *J. Cell. Biol. 25*, no. 1, pt. 2: pp. 79–93.

64. Nayar, G. G., and Sparrow, A. H.
1967. Radiation-induced somatic mutations and the loss of reproductive integrity in *Tradescantia* stamen hairs. *Radiation Botany 7*: 257–267.

65. Newcombe, H. B.
1942. The action of x-rays on the cell. I. The chromosome variable. *J. Genet. 43*: 145–171.

66. Nilan, R. A.
1964. *The Cytology and Genetics of Barley*. Pullman: Washington State University Press, 278 pp.

67. ———, Konzak, C. F., Heiner, R. E., and Froese-Gertzen, E. E.
1964. Chemical mutagenesis in barley. In *Barley Genetics I. Proc. 1st Intern. Barley Genetics Symp.* Wageningen, pp. 35–54.

68. Novick, A., and Sparrow, A. H.
1949. The effects of nitrogen mustard on mitosis in onion root tips. *Heredity 40*: 13–17.

69. Ockey, C. H.
1957. A quantitative comparison between the cytotoxic effects produced by proflavine acetylethyleneimine and TEM on root tips of *Vicia faba*. *J. Genet. 55*: 525–550.

70. Oehlkers, F.
1953. Chromosome breaks influenced by chemicals. *Heredity 6* (Suppl.): 95–105.

71. Osone, K.
1966. Comparison of mutagenic effects of ethyleneimine and ionizing radiations on rice. *Gamma Field Symposia 5*: 53–61.

72. Ramel, C.
1969. Genetic effects of organic mercury compounds.

I. Cytological investigations on *Allium* roots. *Hereditas 61*: 208–230.

73. Rédei, G. P., and Li, S. L.
1969. Effects of x-rays and ethyl methanesulfonate on the chlorophyll *b* locus in the soma and on the thiamine locus in the germline of *Arabidopsis*. *Genetics 61*: 453–459.

74. Salanki, M. S., and Parameswarppa, R.
1968. Colchicine-induced mutants in cotton (*Gossypium hirsutum* L.). *Current Sci.* (India): *37*: 356–357.

75. Sass, J.
1937. Histological and cytological studies of ethyl mercury phosphate poisoning in corn seedlings. *Phytopathology 27*: 95–99.

76. Savage, J. R. K., and Pritchard, M. A.
1969. *Campelia zanonia* (L.) H.B.K.: A new material for the study of radiation-induced chromosomal aberrations. *Radiation Botany 9*: 133–139.

77. Sax, K.
1938. Chromosome aberrations induced by x-rays. *Genetics 23*: 494–516.

78. ———, and Sax, H. J.
1968. Possible mutagenic hazards of some food additives, beverages, and insecticides. *Japan. J. Genetics 43*: 89–94.

79. Siddig, E. A., and Swaminathan, M. S.
1968. Enhanced mutation induction and recovery caused by nitrosoguanidine in *Oryza sativa*. *Indian J. Genet. Plant Breed. 28*: 297–300.

80. Smith, H. H.
1967. Relative biological effectiveness of different types of ionizing radiations: Cytogenetic effects in maize. *Radiation Res. Suppl. 7*: 190–195.

81. ———, and Lotfy, T. A.
1954. Comparative effects of certain chemicals on *Tradescantia* chromosomes as observed at pollen tube mitosis. *Am. J. Botany 41*: 589–593.

82. ———.
1955. Effects of betapropiolactone and ceepryn on chromosomes of *Vicia* and *Allium*. *Am. J. Botany 42*: 750–758.

83. Sparrow, A. H., Baetcke, K. P., Shaver, D. L., and Pond, V.
1968. The relationship of mutation rate per roentgen to DNA content per chromosome and to interphase chromosome volume. *Genetics 59*: 65–78.

84. ———, Cuany, R. L., Miksche, J. P., and Schairer, L. A.
1961. Some factors affecting the responses of plants to acute and chronic radiation exposures. *Radiation Botany 1*: 10–34.

85. ———, and Singleton, W. R.
1953. The use of radiocobalt as a source of gamma rays and some effects of chronic irradiation on growing plants. *Am. Naturalist 87*: 29–48.

86. ———, and Sparrow, R. C.
1949. Treatment of *Trillium erectum* prior to and during mass production of permanent smear preparations. *Stain Tech. 24*: 47–55.

87. ———, Underbrink, A. G., and Sparrow, R. C.
1967. Chromosomes and cellular radiosensitivity. I. The relationship of $D_o$ to chromosome volume and complexity in seventy-nine different organisms. *Radiation Res. 32*: 915–945.

88. Sulouska, K., Lindgren, D., Eriksson, G., and Ehrenberg, L.
1969. The mutagenic effect of low concentrations of ethylene oxide in air. *Hereditas 62*: 264–266.

89. Swanson, C. P., and Merz, T.
1959. Factors influencing the effect of β-propiolactone on chromosomes of *Vicia faba*. *Science 129*: 1364–1365.

90. *Symposium on Chromosome Breakage.*
1953. *Heredity 6* (Suppl.), 315 pp.

91. Taylor, J. H., Haut, W. F., and Tung, J.
1962. Effects of fluorodeoxyuridine on DNA replication, chromosome breakage, and reunion. *Proc. Natl. Acad. Sci. U.S. 48*: 190–198.

92. Veleminsky, J., and Gihner, T.
1968. The mutagenic activity of nitrosamines in *Arabidopsis thaliana*. *Mutation Res. 5*: 429–431.

93. Wagner, J. H., Nawar, M. M., Konzak, C. F., and Nilan, R. A.
1968. The influence of pH on the biological changes induced by ethyleneimine in barley. *Mutation Res. 5*: 57–64.

94. Westergaard, M.
1957. Chemical mutagenesis in relation to the concept of the gene. *Experientia 13*: 224–234.

95. Wettstein, D. von, Gustafsson, Å., and Ehrenberg, L.
1959. Mutationsforschung und Züchtung, *Arbeitsgemeinschaft Forsch. Landes Nordhein-Westfalen 73*: 7–48, 59–60.

96. Wuu, K. D., and Grant, W. F.
1966. Morphological and somatic chromosomal aberrations induced by pesticides in barley. *Can. J. Genet. Cytol. 8*: 481–501.

97. Yamaguchi, H.
1963. The methods for determining the mutation frequency after seed irradiation in rice. *J. Radiation Res. 4*: 97–104.

## Drosophila

While there are a considerable number of mutation tests which are capable of being carried out on *Drosophila*,[1] we refer here primarily to those which in general are simple, economic, rapid and unambiguous in interpretation. The types of tests described below may be run independently, or two or more tests may be carried out on the same group of treated individuals by using special stocks. Whenever quantitative mutation frequencies are required in order to compare, for example, results from different mutagens or different cell stages, then the age of the flies, the breeding periods, and the cell sampling procedures, as well as other physiological and environmental variables, must be rigorously controlled. On the other hand, if only a relative index of mutagenicity is sought, these variables need be less stringently controlled. Detailed procedures will not be presented; however, many of the tests are discussed in general genetics texts or in references listed below.[5]

**Sex-Linked Recessive Lethal Test** Probably the most widely used procedure is the sex-linked recessive lethal test. Either sex may be treated, and mutation frequencies from successive germ cell stages may be obtained. The test requires that two generations be bred; however, since chemical mutagens often produce delayed or mosaic effects, a third generation may be necessary. Large numbers of progeny may be tested. Since a lethal is indicated by the absence of an entire class of flies, the test is objective. Lethals are among the most commonly induced mutations. While the number of gametes analyzable will vary with the number of persons employed, a staff of two or three can screen between 5,000 and 10,000 X-chromosomes a month.

**Two-Generation Reciprocal Translocation Test** The two-generation reciprocal translocation test is in general use in many laboratories. The test is similar to the sex-linked lethal test, requiring single cultures for each $F_1$ individual tested. Screening of $F_2$ progeny is more difficult and time-consuming than for lethals, but the test is an objective and reliable index of chromosome breakage. Meiotic and post-meiotic male germ cells are most effectively studied. Four to six weeks—if retests are carried out—may be required to complete the translocation test.

**One-Generation Sex-Chromosome Loss Test** The sex chromosome loss experiment is a one-generation test which detects either complete or partial loss of the sex chromosomes, the loss resulting primarily from chromosome breakage. The test is useful because either sex may be treated, the phenotypes of the exceptional classes of offspring are readily discernible from the normal progeny, and large numbers of flies may be rapidly screened with each individual representing a treated gamete. It should be possible to examine a minimum of 5,000 progeny from treated gametes per day per investigator. Although many more chromosomes can be tested per manhour by this method than by the recessive lethal method, many mutagens may be more effective at inducing lethals and other point mutations than chromosomal loss.

**Bithorax Test** A second one-generation test of great usefulness in detecting chromosome rearrangements induced in either sex is the bithorax method of Lewis.[3] A conspicuous enhancement of the bithorax phenotype signals a chromosome rearrangement, translocation or inversion, involving chromosome 3, one of the two large autosome pairs of *Drosophila*. Each $F_1$ represents a treated gamete; only the exceptional progeny need be further analysed to verify the transmission and to determine the nature of the change. Probably 4,000 to 8,000 chromosomes can be analysed per week by a single worker. Spontaneous rearrangements are extremely rare.

A third chromosome breakage study applicable only to oöcyte testing involves detachment of attached X-chromosomes.[4, 6] A simple phenotypic difference permits rapid classification of the normal from the exceptional progeny. The spontaneous frequency of detachment is of the order of 1 per 1,500 gametes. Probably 10,000 to 15,000 chromosomes from an array of oöcyte stages can be tested per week per person. The mature oöcyte is perhaps the stage of greatest sensitivity to chromosome damage, as indicated by irradiation studies.

All of the last three mentioned tests need not be counted manually. The investigator can screen for the exceptional flies; the rest can be counted rapidly and accurately by an electronic counter.[2]

References

1. Auerbach, C.
1962. *Mutation*. Edinburgh: Oliver & Boyd.

2. Keighley, G., and Lewis, E. B.
1959. *Drosophila* counter. J. Hered. *50*: 75–77.

3. Lewis, E. B.
1954. The theory and application of a new method of detecting chromosomal rearrangements in *Drosophila melanogaster. Amer. Natural. 88*: 225–239.

4. Muller, H. J., and Herskowitz, I. H.
1954. Concerning the healing of chromosome ends produced by breakage in *Drosophila melanogaster. Amer. Natural. 88*: 177–208.

5. Muller, H. J., and Oster, I. I.
1963. Some mutational techniques in *Drosophila*. In *Methodology in Basic Genetics*, Burdette, W. J., ed., San Francisco: Holden-Day.

6. Parker, D. R.
1954. Radiation induced exchanges in *Drosophila* females. *Proc. Natl. Acad. Sci. U.S. 40*: 795–800.

## Mammalian Methods

### Cytogenetics and Somatic Cell Genetics.

In appraising any method for mutagenicity testing, it is important to be clear as to what we are asking of the test. The advantages and disadvantages of the test system can then be better assessed. With a cytogenetic system, we are seeking for morphological evidence of damage to the genetic material. With this in mind, some of the obvious advantages are that a wide number of species, including human, can be examined by these methods; that the tests can be performed on both in vivo and in vitro systems; that the genetic material is being observed directly; and that the tests can be accomplished relatively rapidly with limited expense. Disadvantages include the facts that the tests need a well-trained investigator for accurate results; that there are possibilities of subjective errors; that procedures need to be standardized, at least within certain limits, in order to ensure that tests are reproducible from laboratory to laboratory; and that there is not complete agreement on definitions and classifications of breaks and gaps, and the various abnormalities. However, the primary disadvantage is this: there is no proof that seeing cytogenetic abnormalities is an absolute indication of mutation. One can visualize a spectrum of damage ranging from such severe damage that the cells die without ever getting into mitosis; to damage such that other cells are made incapable of cell division, but survive in a post-mitotic state, with or without a change in functional proteins; and finally, to gene mutations, in which cells can still go on to divide but have alterations in the functional proteins produced.

Evidence for the first two types of change seems well established. This is, of course, important in our consideration of damage to genetic material, and is a very important consideration in teratogenesis and perhaps in aging. The mutations are less easily confirmed, however. If chromosome breakage is important in mutation, it is because the breaks are an indicator system, since most cells with visible unstable chromosome abnormalities would probably progress to death. Work correlating mutation and chromosome breakage after chemical treatments is in an early stage compared with data available for x-rays. However, there is good correlation between chemicals producing mutations in various systems and those producing chromosome abnormalities.[3] There is almost 100 percent correlation between chromosome aberrations produced in mammalian cells in tissue culture and mutagenic effects, whenever data were available on both effects (Table 3.5). The correlation between mutagenesis and chromosome aberrations in plant root tips is good, but less so than for mammalian cells in tissue culture.

The absolute answer to this question of whether chromosome breaks serve as indicators for gene mutations will probably come from the studies presently starting in somatic cell genetics, when these are correlated with, and studied in conjunction with, cytogenetics. The type of work referred to here deals with the ability of an agent to induce drug resistance which reflects the loss of a functional enzyme in somatic cells as, for example, resistance to BUdR, because the enzyme is no longer incorporated into the cell, due to the loss or modification of the thymidine kinase enzyme. This approach is well exemplified in recent studies. Chinese hamster cells were treated with BUdR, and nutritionally deficient mutants were selected by growing cells in restrictive media in which the nutritionally deficient mutants could not divide; agents that would kill dividing cells were then added. Similarly, selective-culture techniques have been used to isolate L-glutamine

**Table 3.5** Comparison between Chromosome-Breaking and Mutagenic Effects of Chemicals in Plant and Animal Materials

| | Chromosomal Aberrations | | |
|---|---|---|---|
| Compound | Plant Root Tips | Mammalian Cells in Tissue Culture | Mutagenic Effect |
| Adenine | + | + | + |
| 2,6-Diaminopurine | − | + | + |
| Caffeine | + | ± | + |
| 8-Ethoxycaffeine | + | ± | ± |
| Purine riboside | − | + | + |
| Deoxyadenosine | + | + | No data |
| 5-Fluorodeoxyuridine | + | + | No data |
| 5-Bromodeoxyuridine | − | + | + |
| Cytosine arabinoside | − | + | No data |
| Maleic hydrazide | + | − | − |
| Azaserine | + | + | + |
| Streptonigrin | + | + | + |
| Mitomycin C | + | + | + |
| Hydroxylamine | ± | + | + |
| Nitrogen mustard | + | + | + |
| Triethylenemelamine | + | + | + |
| Diepoxybutane | + | + | + |

+marked effect; −no effect; ±effect very low, although just about significant.

Source: Bengt A. Kihlman, *Actions of Chemicals on Dividing Cells*, © 1966. Reprinted by permission of Prentice-Hall, Inc., Englewood Cliffs, N.J.

auxotrophs or 8-azaguanine resistant Chinese hamster cells and to compare the incidence of these mutants in cultures treated with chemical mutagens and control cultures.[1] Human male cells and genes on the X-chromosome have been similarly studied.[2] Loss or deficiency of the enzyme hypoxanthine-guanine phosphoribosyl transferase (HG–PRT) imparts resistance to purine analogues. Since the gene for this enzyme is located on the X-chromosome, the use of male cells permits detection of changes in the single gene. Also, somatic cell genetic systems utilizing isoenzymes via their electrophoretic patterns offer an excellent definitive tool. Changes in these isoenzyme patterns can be proof of mutation in the cultured cells. In addition to confirming the relationship of chromosome breakage to mutation, these somatic-cell systems should, as they are further developed, provide an excellent methodology for mutagenicity testing in their own right.

The importance of demonstrating whether or not chromosome breakage is an indicator for mutations lies in the areas of carcinogenesis, of germ-line mutation with increasing genetic load of the population, and, perhaps in some aspects, of aging.

**Methods** In vitro preparations can be made very

rapidly from tissue cultures and are usually morphologically superior to their in vivo equivalents. Readily available cultures from *Potorous*, designated PTK–1, are exceedingly well suited for cytogenetic studies, in that the chromosomes are large and distinct, and only eleven in number. The Chinese hamster, with twenty-two chromosomes, has many of the same advantages. Of course, there are both human leukocyte cultures and diploid human fibrolast cultures; they have the advantage of human origin. Leukocyte cultures have the additional advantage that their cell cycle is not initiated until phytohemagglutinin is added, so that timing for adding various agents for various portions of the cell cycle can be done with greater precision than possible in many other culture systems.

The in vivo assays offer many of the same advantages of the host-mediated assay which utilizes bacteria (see below). That is, breakdown products and other metabolic products of the test agent, as well as the agent itself, have a chance to produce effects. Bone marrow, spleen, and testes are especially suitable for in vivo preparations, as well as embryo homogenates and tissues.

From all of these materials, both metaphase and anaphase preparations can be made. Metaphase has the advantage of excellent morphologic detail of each chromosome, so that localization to specific chromosomal areas can be accomplished. Anaphase has the advantages that the pretreatment is much reduced; that the rapidity with which anaphase preparations can be read is much greater; and that the experience necessary to become competent in anaphase evaluation is considerably less than for a similar degree of competence with metaphase.

**Classification of Chromosome Breakage** Unfortunately, there is no single classification of breakage, since different criteria of breaks have been used by various authors. This leads to some confusion and redundancy, but, in general, the various classifications are consistent, one with the other. The characteristic that has been used to classify has frequently depended on the type of study underway and the information sought.

One of the main characteristics used to classify chromosome breaks has been whether one or both of the two chromatids of the chromosome are involved in the defect. If both chromatids are involved, the defect is called a chromosome break; if one chromatid is involved it is termed a chromatid break. The factor which determines what type of lesion is produced is whether the chromosome is a single unit or a double unit at the time of the insult which produces the break. This, in turn, is dependent upon the stage of the cell cycle. If a chromosome is in the GI phase of the cycle before DNA synthesis has taken place, it is a single structure. If a break is produced at this time, the break is replicated along with the second chromatid during the S or DNA synthesis period, resulting in a chromosome break. If the breaking insult occurs during G2 or thereafter (after DNA synthesis, when the chromosome is already a dual structure) then a chromatid break is the usual result. During the period of DNA synthesis, a combination of both types of breakage can be found in the same cell, depending on whether the individual chromosome has not yet started, or has finished synthesizing its DNA. It does occasionally happen that an event affects both chromatids after they are a double structure. In this case, the term *isochromatid break* is used, indicating that the lesion was one introduced in chromatids, but that both chromatids were affected at the same point. This is distinguished from chromosome breaks only by the fact that other lesions in the same material are predominantly chromatid breaks.

An additional type of breakage using these criteria has been termed a *delayed isolocus break*.[5] A typical example of this kind of breakage is a secondary constriction in one chromatid with a corresponding break in the isolocus position in the other. In addition to this typical lesion, other chromosomes exhibited every possibility from no more than a secondary constriction in one chromatid to a complete break in both chromatids. It was further considered that a partial defect was produced in the chromosome when it was a single unit, and then this partial defect was reproduced in both chromatids at the isolocus point during DNA synthesis. Mitotic forces and pressures subsequent to this were thought to produce the variety of possible changes at the isolocus spots in the chromatids. An alternative explanation is that this type of breakage occurred during the period of DNA synthesis and affected different chromosomes differently, depending on the state of synthesis of that particular chromosome.

A second important characteristic that has been used in classification of chromosome breaks is dependent on whether or not healing or reunion has occurred. If there is no healing, an open break or defect is the result. This has also been termed a *simple break* and a *terminal deletion*. In this type of breakage, a significant problem arises in distinguishing between a *break*, defined as a "complete discontinuity" between the two chromosome pieces, and a *gap*, defined as an achromatic or unstained area in which chromatin still exists but is difficult to see. Various methods have been used to make this distinction. Some authors insist on displacement of the distal fragment before considering the event a break; others have established an arbitrary distance between the two stained chromosome pieces as the distinguishing factor. We consider any defect separated by at least the width of one chromatid to be a *break*, and anything less than this to be a *gap*. This is admittedly arbitrary, but serves as a useful basis of comparison between experimental and control material. It is fortunate that in most systems, gaps and breaks seem to increase and decrease in parallel, so that the methods described enable valid comparisons between various materials.

When healing or reunion occurs, it is possible for restitution to occur if the broken ends reunite in their original positions. In this case no defect is visible. If the broken ends do not heal in their original positions, a structural rearrangement is the result. These are often further divided into an intrachange if the rearrangement is within a single chromosome, or an interchange if the rearrangement involves more than one chromosome. Both of these can be further divided into symmetrical and asymmetrical defects. A symmetrical defect is one in which no mechanical difficulty results during mitosis, and neither daughter cell is deficient in chromatin material. An asymmetrical intra- or interchange is one in which either mechanical defects arise or the resulting daughter cells are deficient in chromatin material.[4] Another term that is frequently used in a very similar context with *symmetrical* and *asymmetrical* is *stable* and *unstable* rearrangement. The primary factors which determine whether the open or simple type of breakage or the rearrangement will result seem to be whether the cell retains the ability to synthesize protein and/or DNA. If either or both of these processes are interrupted, there is evidence that reunion cannot take place and open breaks will result.

In addition to these classifications, used when the cells under study are examined in metaphase (since metaphase affords greater morphological detail of individual chromosomes due to various pretreatments, including colchicine, hypotonic expansion, and air drying or squashing), it is also possible to score defects in anaphase preparation. Here, none of the previously mentioned pretreatments is used; the cells are merely fixed and stained. The types of anaphase aberration which can be distinguished include an acentric fragment, which is a paired segment of chromatids left at the equator of the cell resulting from a chromosome break; an attached fragment, in which a chromatid fragment is away from the main body of anaphase chromosomes, but which is oriented in line with the chromosomes and which seems to be attached by an attenuated portion, a chromosome bridge, which results from an asymmetrical rearrangement as a dicentric chromosome or an interlocking ring chromosome; and finally, pseudochiasmata, which are thought to result from two chromosomes adhering to each other via stickiness or some other mechanism, and which may very likely not represent true defects.

**Summary** Cytogenetic study certainly seems to be one of the best screening tests for mutagenicity, but should be used in conjunction with other methods. Cytogenetic testing reveals a variety of damage to the genetic material in addition to mutation, as well as a high correlation with mutagenic events when both parameters are tested. It offers in vivo and in vitro methods for a wide variety of species, including the human. When cytogenetic study is a method employed for screening, standardization of procedures, high quality of preparation, and reading of coded slides are essential for best results.

Some difficulties that have arisen in the past, such as the differentiation between gaps and open breaks or the degree of significance of open breaks vs. rearrangements, would seem, on the basis of present information, not to be as big a problem as once imagined. Both gaps and breaks seem to increase in parallel in most of the systems studied up to now. Any method of differentiation between the two, as long as it is standardized (even though arbitrary) is adequate to compare control with experimental material. Similarly, the difference between open breaks and chromosome rearrangements would appear to be whether cellular DNA and/or protein synthesis is inhibited or not. In the absence of DNA and/or protein synthesis, healing is inhibited, and it is the healing that permits rearrangement. Many of the materials, shown to produce only open breaks in acute studies, are seen to progress to chromosomal rearrangements when chronic studies which allow a recovery period are carried out. This difference stresses the need to carry out a portion of the studies in a cytogenetic test system after the test substance has been removed and a recovery period allowed.

References

1. Chu, E. H. Y., and Malling, H. V.
1968. Mammalian cell genetics, II. Chemical induction of specific locus mutations in Chinese hamster cells in vitro. *Proc. Natl. Acad. Sci. U.S. 61*: 1306–1312.

2. De Mars, R.
1969. Personal communication.

3. Kihlman, B. A.
1966. *Actions of chemicals on dividing cells.* Englewood Cliffs, N.J.: Prentice-Hall, 260 pp.

4. Lea, D. E.
1962. *Actions of radiations on living cells.* 2nd ed. New York: Cambridge University Press, 416 pp.

5. Ostegren, G., and Warkonig, T.
1954. True or apparent subchromatid breakage and the induction of labile states in cytological chromosome loci. *Bot. Notiser.* pp. 357–375.

6. Puck, T. T., and Kao, F.
1967. Genetics of somatic mammalian cells. V. Treatment with 5-bromodeoxyuridine and visible light for isolation of nutritionally deficient mutants. *Proc. Natl. Acad. Sci. U.S. 58*: 1227–1234.

## The Host-Mediated Assay

A great deal of recent work in genetics has emphasized the universal nature of the genetic code. Although the level of organization of genetic material in bacteria is different from that in man, there is no necessary basis for assuming that the action of a mutagen will be markedly different. It is essential, however, to define properly the ultimate mutagenic agent occurring in the mammalian host. There are numerous examples of compounds that are not mutagenic in microorganisms, but are converted to active mutagens in animals, and there are many compounds that are active in microorganisms but detoxified in mammalian systems. The host-mediated assay was developed to determine the ability of laboratory mammals either to activate or to detoxify compounds in regard to mutagenic effects.[1–4] Although this is an indirect test, as a microbial indicator is used, it is the only practical method for detecting point mutations in vivo, using micro-organisms as the indicator in mammals.

In this assay,[1] the mammal, during treatment with a potential chemical mutagen, is injected with an indicator microorganism in which mutation frequencies can be measured. It is important to note that mutagen and organism are administered by different routes. After a sufficient time period, the microorganisms are withdrawn from the animal and the induction of mutants is determined. The comparison between the mutagenic action of the compound on the microorganism directly and in the host-mediated assay indicates whether the host can detoxify the compound and whether mutagenic products can be formed as a result of host metabolism. The formation of metabolic products that are mutagenic from dimethylnitrosamine and the plant toxin, cycasin, have been reported using this procedure.

Indicator microorganisms presently being used in this procedure includes the histidine auxotrophs of *Salmonella typhimurium* and *Neurospora crassa*, where scoring for forward mutations is carried out. In the Salmonella system, numerous known auxotrophs are injected intraperitoneally in an animal previously treated with the chemical. After six generations, in approximately three hours, the organisms are recovered from the intraperitoneal cavity and the induction of mutation is determined. The effect in the animal is compared with the effect of the chemical in an in vitro plate assay. In the *Neurospora* system, *conidia* from the *Neurospora dikaryon*, described earlier, can be injected into the peritoneal cavity or subcutaneously in mice or rats. In rats, the *conidia* can also be injected into the testis. After 48 hours, 50 to 70 percent of the *conidia* can be recovered with approximately 50 to 60 percent viability. Since the *conidia* can be kept in the animal an extended period of time, it is possible to do meaningful feeding experiments in which one can test for presence of mutagenic compounds in the diet.[5] After the *conidia* are recovered, they are tested for presence of spontaneous and induced ad–3 mutations.[6] A more ideal indicator which uses a forward mutation system in bacteria is presently being developed. It is probable that newly developed methods for scoring forward and reverse mutations in cultured cells might also be adopted in this procedure.

In addition to flexibility in the selection of indicator organism, almost any laboratory animal can be used; laboratory animals including rats, mice, and hamsters have been used successfully. Not only can we compare mutagenic activity between microorganisms and mammals, but also between different animal species. It should also be possible to demonstrate any correlation between mutagenicity and carcinogenicity in the same or different animals.

The host-mediated assay bridges the gap between

simple microbial tests and the effects of a potential mutagen in mammals. The similarity between mutagenic activity in microorganisms and animals, the ability of the mammal to detoxify mutagens to non-mutagenic agents, and the production of mutagenic metabolites can be determined. Comparisons can be made not only between microorganisms and mammals, but also between different animal species. It might be possible to compare mutagenicity and carcinogenicity in the same system with this procedure. However, the host-mediated assay in no way indicates the effect of DNA repair mechanisms of the host in response to specific chemicals, and is only an indirect measure of mutagenicity in terms of the mammalian host. The host-mediated assay of the future will probably use mammalian, if not human cells, rather than bacteria.

References

1. Gabridge, M. G., and Legator, M. S.
1969. A host-mediated microbial assay for the detection of mutagenic compounds. *Proc. Soc. Exp. Biol.* (N.Y.) *130*: 831–834.

2. ———, Denunzio, A., and Legator, M. S.
1969. Microbial mutagenicity of streptozotocin in animal-mediated assays. *Nature 221*: 68–70.

3. ———.
1969. Cycasin: Detection of associated mutagenic activity in vitro. *Science 163*: 689–691.

4. Gabridge, M. G., Oswald, E. J., and Legator, M. S.
1969. The role of selection in the host-mediated assay for mutagenicity. *Mutation Res.* 7: 117–119.

5. Malling, H. V., and Cosgrove, G. E.
1970. In: Chemische mutagenese bei Sauger und Mensch. Test systeme und ergebnisse. Proceedings of a symposium held October 5–8, 1969, Mainz, Germany. In press.

6. Webber, B. B., and de Serres, F. J.
1965. Induction kinetics and genetic analysis of X-ray induced mutations in the ad-3 region of *Neurospora crassa*. *Proc. Natl. Acad. Sci.* (U.S.) *53*: 430–437.

## Specific Locus Test

The specific locus test enables assay for the induction of recessive lethal mutations in several loci controlling coat-color and morphology in mice.[4, 3] These newly induced mutations can be either chromosome deletions or point mutations. In this test, male mice homozygous for the wild type alleles are treated with the chemical and then, at various times after the administration of the mutagen, are mated with female mice homozygous for the seven recessive alleles. The occurrence of offspring with recessive characteristics among the $F_1$ progeny is indicative of gene mutation. The following compounds, potent mutagens in other organisms, have been tested by the specific locus test: triethylene melamine (TEM),[1] methyl methanesulfonate, ethyl methanesulfonate, n-propyl methanesulfonate, and iso-propyl methane sulfonate.[2] Of these five mutagens, only TEM gave a significant increase in the mutation frequency. The mutation frequency observed in spermatogonia for the four alkyl methanesulfonates combined was only twice the spontaneous rate as determined by Russell in earlier experiments. All four alkyl sulfonates were highly positive in the dominant lethal test. However, the results from the two tests are difficult to compare because the sperm were not treated at comparable stages of development.

The number of animals which must be used to detect a doubling of the spontaneous mutation frequency is so great that the expense of this test makes it absolutely impractical as a general screening test. More serious, however, is the failure, in Russell's experiments, to detect a significant increase in specific locus mutations with four of these five potent chemical mutagens.

In mammals little is known about how point mutations, induced in the germ-line of one generation, are transmitted to the next. In order to evaluate the potential genetic hazards of chemical mutagens to human populations, it seems urgently important that this mechanism be understood.

References

1. Cattanach, B. M.
1966. Chemically-induced mutations in mice. *Mutation Res. 3*: 346–353.

2. Ehling, U. H., and Russell, W. L.
1969. Induction of specific locus mutations by alkyl methanesulfonates in male mice. *Genetics 61*: 14–15.

3. Lyon, M. F., and Morris, T.
1966. Mutation rates at a new set of specific loci in the mouse. *Genet. Res. Camb. 7*: 12–17.

4. Russell, W. L.
1951. X-ray induced mutations in mice. *Cold Spring Harbor Symp. Quant. Biol. 16*: 327–366.

## Dominant Lethal Test

Dominant lethal mutants are convenient indicators of major genetic damage which have been used in mammals for measuring effects of x-rays,[1] and, more recently, of chemical mutagens.[2,3,7,8,10,11,13,15,23] Data on induction of dominant lethal mutants in mammals may be appropriately extrapolated to man, especially as most recognizable human mutations are due to dominant autosomal traits.[21] The genetic basis for dominant lethality is the induction of chromosomal damage and rearrangements, such as translocations and aneuploidies, resulting in non-viable zygotes. Evidence for zygote lethality induced in mammals by x-rays and by chemical mutagens has been obtained embryologically[16,26,27] and cytogenetically.[4,15,17,24] Additional evidence for the genetic basis of dominant lethality is derived from the associated induction of sterility and heritable semisterility in $F_1$ progeny of males exposed to x-irradiation,[19,26] and to chemical mutagens.[5,14] Translocations have been cytologically demonstrated in such semisterile lines in mice[7,18,25] and in hamsters.[20]

The induction of dominant lethal mutations in animals can be assayed, with a high degree of sensitivity and practicality, following acute, subacute, or chronic administration of test materials, either orally or by any parenteral route, including the respiratory. For these reasons, it is feasible to integrate such tests in the scope of routine toxicological practice.[9] Following drug administration to male rodents, the animals are mated sequentially with groups of untreated females over the duration of the spermatogenic cycle. For mice, the entire duration of spermatogenesis is approximately 42 days comprising the following stages: spermatogonial mitoses, 6 days; spermatocytes, 14 days; spermatids, 9 days; testicular sperm, 5.5 days; and epididymal sperm, 7.5 days.[1] Thus, matings within 3 weeks after single drug administration represent samplings of sperm exposed during post-meiotic stages, and matings from 4 to 8 weeks later represent samplings of sperm exposed during pre-meiotic and stem-cell stages.

The classical form of the dominant lethal assay involves autopsy of females approximately 13 days following timed matings, as determined by vaginal plugs in mice and vaginal cytology in rats, and enumeration of corpora lutea and of total implants, as comprised by living fetuses, late fetal deaths, and early fetal deaths. The test has been considerably modified and simplified, and hence made more suitable for routine practice, by sacrificing the females at a fixed time, e.g., 13 days in mice, following the midweek of their caging and presumptive mating.[13]

Additionally, this allows determination of effects of drugs on pregnancy rates. Similarly, corpora lutea counts, which are laborious and inaccurate in mice and afford a measure of total fertilized zygotes, can be omitted. Numbers of total implants in test animals can be related to those in controls, thus affording a simple measure of preimplantation losses. Using such modified procedures, together with computerized data handling, large numbers of test agents can be simply and rapidly tested for mutagenic activity.[10,11,13] Dominant lethal mutations are measured directly by enumeration of early fetal deaths and indirectly by preimplantation losses, as measured by reduction in the number of total implants in test, as compared with control, females.

It has been suggested that the assay can also be conducted by administering test agents to female mice, either before or in early pregnancy.[15,22] However, such results are complicated not only by the possibility of teratogenic and other effects, but also by the fact that only postmeiotic germinal stages can be tested with available practical techniques.

Results are best expressed as early fetal deaths per pregnant female, rather than the more conventional mutagenic index, early fetal deaths × 100 per total implants; the latter index can be markedly altered by variation in the number of total implants.[10] Preimplantation losses offer a presumptive index of mutagenic effects, but there is no precise parallelism between preimplantation losses and early fetal deaths. These should be regarded as concomitant and not alternate parameters.[10] Furthermore, the use of the mutagenic index presupposes that the number of early deaths is proportional to the number of implants regardless of preimplantation losses; this anticipates that absolute number of early deaths is lower in those animals with reduced numbers of total implants. This has been shown experimentally not to be so.[10] Finally, an additional disadvantage of such ratios as measures of mutagenic effect is that their variability is high; both numerator and denominator are contributory, and estimates of standard deviation, therefore, are complex.

Preimplantation losses, early fetal deaths, sterility, and semisterility constitute a spectrum of adverse genetic effects, of which early fetal deaths clearly afford the most convenient and quantitatively unequivocal parameter of mutagenicity.[10,17]

Using these techniques, a wide range of chemicals to which man is exposed in the totality of the environment, including pesticides, food additives, drugs, and air and water pollutants, have been tested for mutagenicity in mice.[10, 11, 13] Additionally, detailed dose-response studies with the aziridine alkylating agents, TEPA and METEPA, which have been used as chemosterilant pesticides, have revealed mutagenic thresholds in the region of 0.04 mg/kg and 1.4 mg/kg, respectively, following acute single parenteral administration in mice.[10]

These techniques are also ideally suited for the study of synergistic or antagonistic effects on mutagenesis. Caffeine, for example, has been shown neither to induce dominant lethal mutations nor to synergize the mutagenic effects of x-rays or of alkylating agents.[12]

References

1. Bateman, A. J.
1958. Mutagenic sensitivity of maturing germ cells in the male mouse. *Heredity 12*: 213–232.

2. ———.
1960. The induction of dominant lethal mutations in rats and mice with triethylenemelamine (TEM). *Genet. Res. Camb. 1*: 381–392.

3. ———.
1966. Testing chemicals for mutagenicity in a mammal. *Nature 210*: 205–206.

4. ———.
1969. Personal communication.

5. Cattanach, B. M.
1964. A genetical approach to the effects of radiomimetic chemicals on fertility in mice. In *Effects of Ionizing Radiation on the Reproductive System*, Carlson, W. D. and Gassner, F. X., eds. New York: Macmillan, pp. 415–426.

6. ———, and Edwards, R. G.
1958. The effects of triethylenemelamine on the fertility of male mice. *Proc. Royal Soc. Edinburgh B. 67*: 54–64.

7. ———, Pollard, C. E., and Isaacson, J. H.
1968. Ethylmethanesulfonate-induced chromosome breakage in the mouse. *Mutation Res. 6*: 297–307.

8. Ehling, U. H., Cumming, R. B., and Malling, H. V.
1968. Induction of dominant lethal mutations by alkylating agents in male mice. *Mutation Res. 5*: 417–428.

9. Epstein, S. S.
1969. A *Catch-All* Toxicological Screen. *Experientia 25*: 617.

10. ———, Arnold, E., Steinberg, K., Mackintosh, D., Shafner, H., and Bishop, Y.
1970. Mutagenic and antifertility effects of TEPA and METEPA in mice. *Toxicol. Appl. Pharmacol.*, 17: 23–40.

11. ———, Bass, W., Arnold, E., and Bishop, Y.
1970. The mutagenicity of trimethyl phosphate in mice. *Science 168*: 584–586.

12. ———.
1970. The failure of caffeine to induce mutagenic effects or to synertize the effects of known mutagens in mice. *Fd. Cosmet. Toxicol.*, 8: 381–401.

13. Epstein, S. S., and Shafner, H.
1968. Chemical mutagens in the human environment. *Nature* 219: 385–387.

14. Falconer, D. S., Slizynski, B. M., and Auerbach, C.
1952. Genetical effects of nitrogen mustard in the house mouse. *J. Genet. 51*: 81–88.

15. Generoso, W. M.
1969. Chemical induction of dominant lethals in female mice. *Genetics 61*: 461–470.

16. Hertwig, P.
1940. Vererbarer semisterilität bei Mäusen nach Röntgenbestrahlung, verursacht durch reciproke Translokationen. *Z. Indukt. Abstamm.-U. Vererb. 79*: 1–27.

17. Joshi, S., Arnold, E., Bishop, Y., and Epstein, S. S.
1970. *Genetics, 65*: 483–494.

18. Koller, P. C.
1944. Segmental interchange in mice. *Genetics 29*: 247–263.

19. ———, and Auerbach, C.
1941. Chromosome breakage and sterility in the mouse. *Nature 148*: 501–502.

20. Lavappa, K. S. and Yerganian, G.
1969. Personal communication.

21. Report of the United Nations Scientific Committee on the Effects of Atomic Radiation.
1966. N. Y. United Nations, pp. 99.

22. Röhrborn, G.
1967. Dominant Lethals in young female mice. *Mutation Res. 4*: 229–231.

23. ———.
1968. Mutagenicity tests in mice. *Humangenetik 6*: 345–361.

24. Russel, L. B., and Russel, W. L.
1954. Pathways of radiation effects in the mother and the embryo. In *Cold Spring Harbor Symp. on Quant. Biol. 19*: 50–59.

25. Slizynski, B. M.
1952. Pachytene analysis of Snell's T(5:8) *a* translocation in the mouse. *J. Genet. 50*: 507–510.

26. Snell, G. D., Bodemann, E., and Hollander, W.
1934. A translocation in the house mouse and its effect on development. *J. Exp. Zool. 67*: 93–104.

27. ———, and Picken, D. I.
1935. Abnormal development in the mouse caused by chromosome unbalance. *J. Genet. 31*: 213–235.

## Population Monitoring

Whatever may be the system of testing potential pesticides before their registration and use, it can never be perfect. There is always the possibility that some will not be revealed as mutagenic by any of the test systems employed and yet present a genetic risk to man. An example might be a pesticide that is not mutagenic per se, but which is specifically converted by the human body into a metabolite that is strongly mutagenic. If such a pesticide or other compound were widely used, we could be doing great harm to our descendants and never discover this fact until the damage had already occurred. As we have emphasized, genetic damage is irreversible as far as is now known.

Is there any possibility of setting up a system to detect such a genetic emergency if it should occur? The task would be enormously difficult, for many reasons already mentioned. For one thing, the damage caused by mutations occurs in future generations, not this one, so the effect would not be observed for some time. In the second place, the effect might be spread out over many generations so that an enormous total effect would still be small enough, in the first generation, not to be noticed. Finally, the kinds of effects produced by mutations are not unique, so if there were, for example, an increased disease or death rate it would be very difficult to be sure that this were due to mutation and not some other cause.

We have to accept, for the present, the fact that any feasible system of monitoring the human population is likely to detect only a very gross effect. But, of course, that is what is most to be feared. So there may be merit in setting up a system that would detect, at the earliest possible date, a really large increase in the mutation rate—say an increase of several times—if this is occurring. Can such a system be made workable, and not prohibitively expensive? We do not have these answers now, but we would like to suggest a few possibilities which might merit further consideration. The problem of detecting environmental mutagens must, of course, be considered in a wider context than for pesticides alone.

A direct search for an increased rate of occurrence of malformations and diseases of genetic origin would necessarily involve a delay of at least 9 months, as a mutation that occurs in the parent will be seen only after the child is born. In the future, intrauterine tests may become feasible; at present, the techniques are not adapted to the wide-scale application that would be necessary if a general rise in mutation rate were to be detected. It might be possible to select certain traits that would be the most efficient indicators of an increased mutation rate. Such indicator traits would have to be (a) dominant, so that the trait shows up in the next generation after the mutation occurs; (b) present at the time of birth, or shortly after, so that there is minimal delay in the discovery; (c) conspicuous, so that the traits would be unambiguously and easily detected by those attending the birth; (d) of a unique appearance not mimicked by other traits not of mutational origin; (e) of such a nature that it is easy to distinguish new mutants from those cases where the parent had the trait and transmitted it to the child. This latter point could be ensured by having the trait of such a nature as to lead to sterility so that every case is a new mutation.

The number of traits that meets these exacting criteria is very limited; we know of none that does absolutely. But there are probably several that come somewhere near. It is difficult to recommend a specific set, but we think the possibility ought to be investigated further. As an alternative to choosing traits so conspicuous and characteristic that they would always be recognized, one might have those attending the birth simply report all instances where the child is abnormal and then have a staff of specialists in congenital anomalies visit each case. The proportion of births with obvious anomalies is in the vicinity of 1 or 2 percent. By examination of a tiny fraction of all children born, the specialist would have an excellent chance of selecting among these those with defects likely to be mutational in origin.

The success of such a system would depend not only on getting good observations at the source, but also on a system of prompt reporting and data analysis, so that any trend could be detected promptly. If an increase is detected, one could hope to identify the cause by such factors as geographical patterns.

Some geneticists think that monitoring gross abnormalities, such as those discussed here is too crude. They suggest that it would be better to use refined chemical procedures that can detect changes in the proteins that are the immediate gene products. At present such tests are very expensive, but there is no question that ultimately such approaches will be possible. The rough and ready and the refined methods are not mutally exclusive. Both have their advantages. We think it is likely that, as our chemical environment becomes increasingly more complicated, more and more elaborate systems of monitoring will be necessary.

The cost of genetic monitoring such as has been discussed would be very great. It could probably be justified only if it were a part of a general system of monitoring for other environmental factors. It would be natural to couple a mutation-detecting system with a search for new teratogens in the environment. Our memory of the thalidomide disaster is a reminder of the need to have a system that will reveal, as promptly as possible, any agent that is causing physical abnormalities and disease, whether from increased mutation or any other cause.

Another possibility for monitoring is to study the human population, as before, but instead of looking at the next generation, look at this generation for changes that might foreshadow such changes in the future. If mutations are induced in the germ cells, they are also induced, in all probability, in other body cells. Therefore, a sensitive system of monitoring mutation rates in the blood cells could give a much quicker indication of an environmental change. Such tests could be both chemical tests for altered gene functions and cytological observations for chromosome aberrations.

## Conclusions

A number of procedures is presently available in mammals, the majority of recent origin, that can be used to determine the mutagenic activity of chemicals. Our ability to characterize mutagenic agents no longer depends exclusively on nonmammalian systems, such as *Drosophila*, bacteriophage, micro-organisms, and in vitro cell culture, although these procedures should be considered as ancillary to the available mammalian tests. The mammalian tests which should be considered as the definitive basis for evaluating potentially mutagenic agents are the host-mediated assay, in vivo cytogenetic studies, and the dominant lethal test. *These procedures are at least as relevant to man as any other animal procedure presently used in the field of toxicology.* They are also practical. The dominant lethal test can be concluded in less than three months, whereas cytogenetic studies and the host-mediated assay can be carried out in a few weeks. The cost of these tests is considerably less than that of many procedures currently used in chronic toxicity testing. It is anticipated that a testing protocol, relying on both the outlined mammalian tests and the ancillary procedures, should detect the majority of mutagenic chemicals. The possibility of integrating mammalian mutagenicity tests with other toxicological procedures has been recently recommended.[1]

Since the mammalian procedures presently recommended are of comparatively recent origin, continued improvements in these techniques can be anticipated; such improvements are of course desirable for any toxicological procedure. Most important is the need for inexpensive and sensitive tests that can detect point mutations in mammals. Particularly promising in this connection is the development of systems in mice that combine genetically marked chromosomes with crossover-suppressing inversions, which can be used to detect recessive lethal mutations.

Reference

1. Epstein, S. S.
1969. A *catch-all* toxicological screen. *Experientia* *25*: 703–704.

# 4 A Recommended Program for Mutagenesis Testing

There are several bases for selecting pesticides which are likely to be mutagenic and which need most to be tested. Clearly, it is important to test those used on a wide scale to which large numbers of humans are exposed. In this context, pesticides used in the home may be more important than those used in areas away from humans and human crops.

It is possible to make some predictions as to which pesticides are most likely to be mutagenic. Substances which are known to be teratogenic or carcinogenic, or which interfere with reproductive functions, are also often mutagenic. Chemical structure can sometimes be a useful guide to predicting possible mutagenicity. For example, many alkylating chemosterilants could have been predicted to be mutagenic in advance of actual tests.

If priorities are needed, at the top of the list are those pesticides produced in the largest amounts, with the greatest emphasis on those used domestically and on food crops. Particular attention must be directed to domestic exposure, by inhalation, of pesticide aerosols and vaporizing pesticide strips i.e., Dichlorvos. The possibility exists that this may represent a major source of hitherto unsuspected human exposure. It should also be stressed that labile pesticides (such as Captan) with a half-life of 10 seconds in serum pose as potentially serious a mutagenic hazard as do persistent pesticides.

There are only about four hundred chemicals commonly incorporated in current pesticide formulations.[1] It is feasible to test all of them, using mammalian and ancillary procedures recommended in this monograph, within a reasonable period of time—say, a year.

Although we cannot foresee all contingencies, we recommend the following as a generally feasible protocol:

(a) Test all pesticides now used in the following: (1) three mammalian systems, the dominant lethal, host-mediated, and in vivo cytogenetic, by appropriate routes of administration which reflect human exposure, and also parenterally and at high-dose levels, such as maximal tolerated doses; (2) ancillary microbial systems, preferably those detecting both single nucleotide changes and effects involving more than one gene. The precision of testing, both in mammalian and ancillary systems, should be such that a doubling of the control level of mutation would be statistically significant at the 5 percent level. A pesticide is regarded as nonmutagenic if none of the tests yields results significantly different from its control. If one or more of the three mammalian tests shows a significant effect, the test is regarded as positive. If only the microbial test is positive, more detailed mammalian tests are indicated.

(b) If the compound is inactive in all systems, then it is tentatively assumed to be safe. If the compound is widely used or if, for any reason, there is the possibility of extensive human exposure, it is advisable that more extensive tests be made. Those compounds which have the greatest chance of having an effect on man should be additionally tested, to take into account problems of possible interactions and duration and rate of exposure.

(c) If the compound is mutagenic, a reasoned decision must be made as to whether the benefit is great enough to warrant further detailed evaluation, with appropriate interim restrictions on use, or whether its use must be disallowed forthwith.

(d) No new pesticide should be registered until tested for mutagenicity, in addition to other standard toxicological procedures.

(e) The testing procedures recommended above must be constantly updated and improved to reflect new techniques and new data. We therefore recommend further that a group of disinterested and scientifically competent persons be assigned the problem of continuously reviewing the whole question of pesticide mutagenesis and test systems to be employed.

Reference

1. Neumeyer, J., Gibbons, D., and Trask, H. 1969. Pesticides. *Chemical Week*, Apr. 12 and 26, pp. 38–68.

# 5 Structure-Activity Relations of Pesticides

Most pesticides have not been designed with the aim of attacking the hereditary material of cells, but their activity has been discovered by chance or they have been designed as analogs of known metabolic inhibitors of activators.[1,7,10] If their assumed metabolic effect is actually responsible for their pesticide action, it should be possible to replace mutagenic by nonmutagenic compounds, except when the type of inhibition itself causes mutations, such as inhibition of DNA synthesis. The known or suspected major modes of action of pesticides are listed here:

1. Plants (herbicides, fungicides)

(a) Inhibition of photosynthesis (triazines; substituted ureas; carbamates; bipyridylium quarternary salts)

(b) Inhibition of oxidative phosphorylation (dinitrophenol analogs, such as toluidines; carbamates)

(c) Hormone (auxin) analogs (2,4–D; 2,4,5–T; benzoic acid analogs; possibly maleic hydrazide)

(d) Inhibition of pantothenate synthesis (chlorinated aliphatic hydrocarbons)

(e) Inhibition of porphyrin, hence of chlorophyll synthesis (Amitrole, which also inhibits purine synthesis)

(f) Unknown causes (metals, sulfur)

2. Animals (insecticides, nematocides; most are nerve poisons)
(a) Inhibition of acetylcholinesterase (organophosphates, carbamates)
(b) Inhibition of neuromuscular junction (nicotinoids)
(c) Neurotoxication with only partially known causes (chlorinated or brominated hydrocarbons, pyrethroids)

Exceptions to the general rule are chemosterilants designed to produce dominant lethal mutations in insects, giving rise to nonviable offspring.[2, 8, 9] In the U.S., chemosterilants are not registered for use as pesticides. However, the Entomology Division of the USDA is currently conducting experimental field studies, in some of which the possibility of human exposure cannot be excluded. More alarming is evidence of active commercial interest in chemosterilants in Japan, where extensive field tests are now in progress.[5] It should be emphasized that chemosterilants must never be employed outside the laboratory except under rigorously supervised conditions.

Most reactions which alter the hereditary information in a cell seem to be caused by a chemical or enzymic attack on DNA itself. The major exception to this rule is the effect of colchicine, which inhibits spindle formation and causes the production of polyploids. Agents which alter DNA itself produce either "mutagenic" or "inactivating" DNA alterations. Mutagenic DNA alterations are minor alterations of the DNA bases which do not prevent DNA duplication, but which cause, occasionally or always, a change in the base sequence of DNA. Such DNA alterations are induced by base analogs (2-aminopurine, 5-bromouracil), which are incorporated into DNA, or by the chemical alterations of DNA bases, such as the deamination of adenine or cytosine by nitrous acid, the hydroxylamination of cytosine by hydroxylamine, the alkylation of guanine by alkylating agents, or by the intercalation of acridine dyes between DNA bases. Mutagenic DNA alterations give rise to point mutations.

In contrast, inactivating DNA alterations have more drastic effects on DNA, since they inhibit the duplication of DNA across the altered side. Such alterations arise when a DNA base is removed or the DNA backbone is broken as a consequence of treatment by alkylating agents, radical-producing agents, or base analogs which inhibit the duplication of DNA. Many inactivating DNA alterations can be repaired by special cellular enzymes. Different organisms differ in the extent and specificity of repair mechanisms. DNA alterations that have not been repaired lead to chromosome breaks which are usually lethal to the cell. If more than one chromosome break occurs within a cell, large, heritable chromosome aberrations can be produced e.g., deletions, translocations or inversions. Most, if not all, agents which induce inactivating DNA alterations or chromosome breaks in vivo have also been found to induce mutations, cancer, and teratogenic effects, when tested in the appropriate system.[3, 4, 6]

Most compounds affecting DNA produce both mutagenic and inactivating DNA alterations, but the relative frequency of these two effects may differ up to one million times for different compounds.[3, 4] A mutagenic test system which is very sensitive for one compound, e.g., transition type point mutations, may reveal no mutations with another compound that produces only other types of mutations, e.g., large chromosome alterations. There is also no correlation between toxic and mutagenic effects, because some highly mutagenic compounds, e.g., certain base analogs, are barely toxic, whereas some highly toxic compounds, such as cyanide, are hardly mutagenic.

On the basis of theoretical extrapolation from available data, pesticides may be classified in the following three major groups by chemical structure.

A. Compounds known to alter DNA directly in some biological system or compounds having chemical structures that are known to alter DNA: alkylating agents, radical-producing agents, inhibitors of DNA synthesis. Irrespective of any further tests, extreme care with respect to human exposure is recommended.

B. Compounds which, by their structure, may possibly affect DNA either directly or after enzymic activation into reactive metabolites or derivatives. Among these are mercury compounds, some of which are known mutagens, and carbamates, some of which may be converted by plant and animal systems into N-hydroxycarbamates that are known to produce chromosome breaks. Due to problems of uptake, enzymic activation or inactivation, and accumulation, it is not possible to make any safe prediction on the mutagenicity of these compounds in mammals. Thorough testing is of course necessary irrespective of attempted chemical predictions.

C. Compounds not suspected to produce genetic

alterations because their chemical reactivity with DNA or their mutagenicity has not been tested and their structure does not suggest such activity. Among the unsuspected chemical structures are cyclopropane rings, as found in pyrethrins, triazines, 2,4–D and other auxin analogs, and those chlorinated hydrocarbons which do not belong to group A or B (e.g., DDT). Nevertheless, our general ignorance concerning metabolic conversions makes it desirable that all these compounds also be tested for mutagenicity. Some triazines, e.g., aziridines, also have alkylating groups and for that reason belong in Group A.

Chemical structures known or suspected to affect DNA and pesticides having these structures are indicated as follows:

A. Compounds having chemical structures that are known to affect DNA.

Any compound having such a structure should be proven to be harmless before humans are exposed to it.

**(a) Alkylating agents** These induce both point mutations (transitions) and large chromosome alterations.

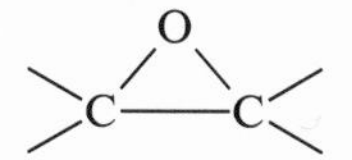


Epoxides (ethylene oxide; endrin; dieldrin)

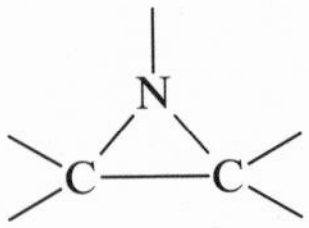

Ethyleneimines (aziridines, such as Apholate; TEPA; thio-TEPA; METEPA)

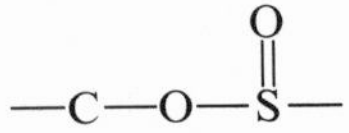

Sulphates (Aramite)

Certain bromides and chlorides (bromomethane; bromopropane; dichloroethane; ethylene dibromide; propargylbromide)

**(b) Radical producing agents** These induce large chromosome alterations but not point mutations.

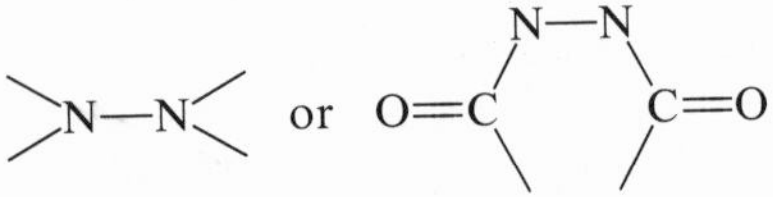

Hydrazines or hydrazides (maleic hydrazide, known to break chromosomes)

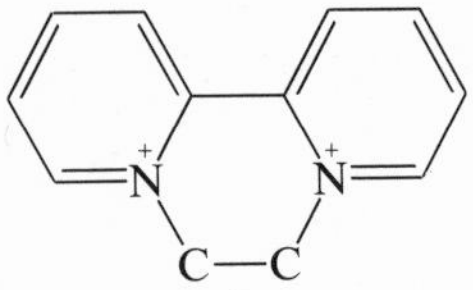

Bipyridylium quarternary salt (Diquat; paraquat). Known to produce radicals and to kill plants in the presence of oxygen (presumably via peroxide radicals affecting DNA)

**(c) Base analogs**

5-Fluorouracil; fluoroörotic acid. Both inhibit DNA synthesis.

B. Compounds suspected to affect DNA or to be activated enzymatically.
These pesticides must be tested for their mutagenicity in higher organisms.

**(a) Unsaturated rings with -OH or SH groups** Some phenols and cresols are known to produce chromosome breaks, perhaps due to radical formation in presence of oxidizing groups.

Ioxynil, niacide; orthophenylphenol; PCP

**(b) Carbamates and Thiocarbamates**

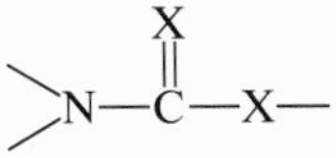

$$>N-\overset{\overset{X}{\|}}{C}-X-$$

X = O or S

Barban; Bux Ten; carbaryl; carbofuran; Carzol; CDAA; dimetilan; metham; Mobam; propachlor; 2,3,6 TBA; Temik; thiram; vapam; ziram
Ethylcarbamate and several other carbamates are well known to produce chromosome breaks, cancer, and teratogenic effects. They do not affect DNA directly, but their enzymic products, such as N–hydroxycarbamates and other intermediates, which in turn produce radicals, cause inactivating DNA alterations; thus, chromosome breaks and large chromosome alterations, but not point mutations, can be induced. Whether or not a particular carbamate is mutagenic depends, therefore, on the presence of N–hydroxylating enzymes in the cells.

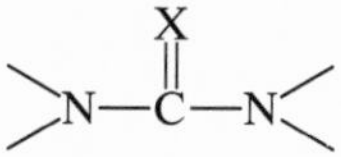

$X = O$ or $S$

### (c) Ureas

Chloroxuron; DCU, diuron; fenuron; Linuoron; norea. These compounds also produce inactivating DNA alterations if they are activated (N–hydroxylation)

### (d) Compounds having 2 or 3 nitrogens connected

If they are known to produce alkylating agents (certain nitroso compounds) or radicals, they should be placed in Group A.

Triazoles (amitrole)

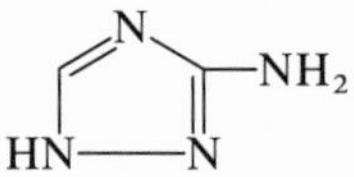

Benzotriazines (azinphosethyl; azinphosmethyl)

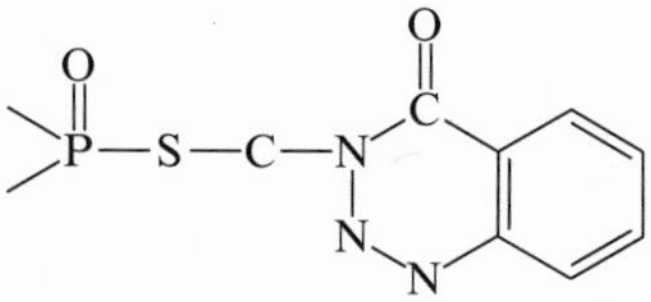

Diazo compounds (Dexon)

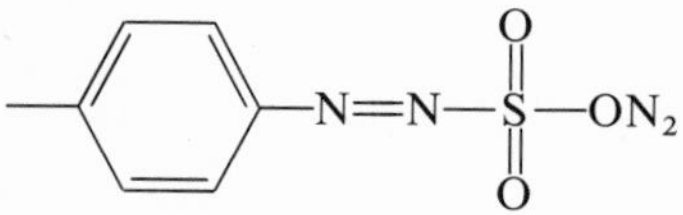

—Hg—

(e) **Mercurials** Some mercurials are known mutagens. Elcide; Emmi; ethylmercury chloride; Panogen; Ceresan M; Semesan Bel.

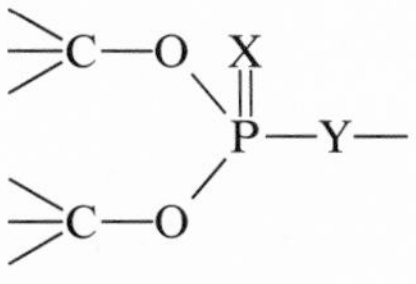

(f) **Organophosphates** These compounds are phosphate triesters which are labile to hydrolysis. Although not tested, some of them may react with DNA by transalkylation.

>C—O, X, P—Y—, >C—O

X = O or S

Y = O—, S—, or N<

Parathion; Methylparathion; malathion; demeton; disulfotone; dimethoate; phorate; phosphamidon; Cyolane; dimefox; Monitor; Ruelene.

(g) **Other reactive compounds**
Acrolein; allyl alcohol; acrylonitrile; (known to react with 4-uridine, inosine, t-RNA)

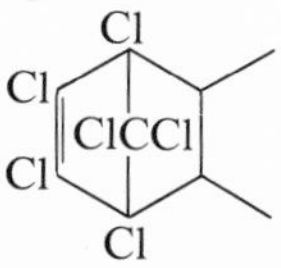

Cl, Cl, ClCCl, Cl, Cl

(h) **Chlorinated cyclodienes** having an unsaturated group in a ring attached to the chlorinated one. These compounds are known to be converted into epoxides that are found in body fat.

(Aldrin; Isodrin; heptachlor)

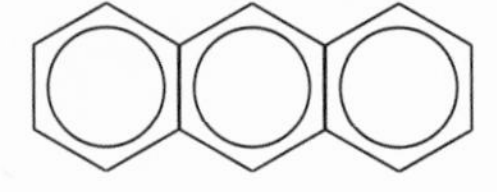

(i) **Base analogs** Some of these compounds may possibly inhibit DNA synthesis.
(Benlate; isocil; lenacil; terbacil).

(j) **Arsenates and cacodylic acids** inhibit phosphorylation

(k) **Potential intercalating compounds** Some of these compounds may act similarly to acridine and intercalate between DNA bases.

(Anthraquinone; Morestan; phenothiazine)

(l) **Certain antibiotics**
(Griseofulvin)

References

1. Brian, R. C.
1964. The classification of herbicides and types of toxicity. In *The Physiology and Biochemistry of Herbicides*, ed. Audus, L. J. part 1. New York: Academic Press. pp. 1–37.

2. Epstein, S. S., Arnold, E., Steinberg, K., Mackintosh, D., Shafner, H., and Bishop, Y.
1970. Mutagenic and antifertility effects of TEPA and METEPA in mice. *Toxicol. Appl. Pharmacol.*, 17: 23–40.

3. Freese, E.
1963. Molecular mechanism of mutations. In *Molecular Genetics*, ed. Taylor, J. H. part 1. New York: Academic Press. pp. 207–269.

4. ———, and Freese, E. B.
1966. Mutagenic and inactivating DNA alterations. *Radiation Res. 6* (Suppl.): 97–140.

5. Japan Insect Sterilant Association, vol. *3*, 1966–1967.

6. Kalter, H.
1968. Teratology of the Central Nervous System. Chicago: University of Chicago Press.

7. Kihlman, B. A.
1966. *Actions of Chemicals on Dividing Cells.* Englewood Cliffs, N.J.: Prentice-Hall, Inc.

8. Kilgore, W. W.
1967. Chemosterilants. In *Pest Control*, ed. Kilgore, W. W., and Doutt, R. L., part 1. New York: Academic Press, pp. 197–239.

9. Loveless, A.
1966. *Genetic and Allied Effects of Alkylating Agents.* Butterworths, London.

10. O'Brien, R. D.
1967. Insecticides: Action and metabolism. New York: Academic Press.

# 6 Usage Patterns of Pesticides

The expected increased need for pesticides in the next five to ten years may result in the greater use of currently registered chemicals. New products are also being rapidly developed and will no doubt replace a number of widely used pesticides. The enclosed summary lists the major categories of pesticides that predominate on the market.

## Herbicides

A. Pre-emergence herbicides, such as cacodylic acid (an arsenical), atrazine (triazine derivative), trifluralin (a dinitrotoluidine), and Amiben (benzoic acid derivative) lead the use of herbicides, and are expected to continue to grow over the next several years.

B. Post-emergence herbicides, such as phenoxyacetic acid derivatives (i.e., 2,4–D, 2,4,5–T), represent a major-use category.

## Insecticides

For years DDT and other chlorinated hydrocarbons dominated the insecticide market. However, due to their persistence, ecological effects, and other hazards i.e., carcinogenicity, they probably will be phased out of agricultural use in the U.S. in the early future. Malathion, a phosphate insecticide, will be likely to increase in usage over the next several years, and represents a major class of insecticides in use. Carbaryl, a carbamate insecticide, is expected to continue to grow in use and is now a major insecticide. Similarly, the systemic insecticides, such as carbofuran, dimethoate, disulfoton, methyl-demeton, phorate, and phosphamidon, are major products in use today.

Based on U.S. sales at manufacturing levels, the following materials can be considered the most widely used in agriculture:

Herbicides: atrazine; trifluralin; Amiben and related products; 2,4–D; 2,4,5–T; propachlor, CDAA, and related products; picloram; paraquat; dicamba; nitralin.

Insecticides: carbaryl; malathion; aldrin; Diazinon; toxaphene; DDT; Methyl parathion; Parathion; azinphosmethyl; chlordane; heptachlor; disulfoton; phorate; Dicofol; Bux Ten; endrin.

Fungicides: dithiocarbamates; Captan; Pentachlorophenol; Dodine.

Fumigants: methyl bromide.

Of these agriculturally important materials, several should be given extra consideration because of their widespread usage in household formulations. Among these are chlordane, DDT and its analogs, BHC and other chlorinated hydrocarbons, including the cyclodienes, and certain organophosphates such as Malathion. In addition, a number of other products are used particularly for household purposes, especially in aerosols and vaporizing strips, and should consequently be considered as priority compounds for mutagenic evaluation. Included in this group are the pyrethroids and their synergists, i.e., piperonyl butoxide, dichlorvos, and such commonly applied insect repellants as diethyltoluamide and ethylhexanediol. Some of these materials are particularly important because human exposure occurs largely through inhalation via aerosol sprays or vaporization, or through skin absorption, quite apart from the more widely recognized oral route.

New methods of pest control, based on biological, physical and highly restricted chemical techniques, have been recently evolved.[1] These methods are effective, pose no potential human hazards such as carcinogenicity, teratogenicity, or mutagenicity, and produce no undesirable disturbances of the ecosystem. It is hoped that much greater interest and attention will be directed to such alternate methods of pest control in the immediate future. The restricted exploitation of alternate methods, hitherto, has reflected their limited economic incentives.

Reference

1. Kilgore, W. W., and Doutt, R. L. (eds.). 1967. *Pest Control: Biological, Physical, and Selected Chemical Methods*. New York and London: Academic Press.

# 7 Literature on Mutagenicity of Pesticides

Approximately four hundred chemicals are now used in the control of weeds, insects, nematodes, rodents, and plant diseases.[5] In the literature search summarized in the Appendix, more than five hundred published papers on the mutagenicity of pesticides have been located. Many of these papers refer to pesticides which are not currently in common use in the United States, but which, in some cases, are used elsewhere. Therefore, they represent a potential hazard to the population of the United States, either as contaminants in imported foodstuffs, or through future registration in the United States. In this preliminary manual literature search, we located 42 papers referring to mutagenic testing of some compounds in a recent listing.[5] In total, 31 of the 32 compounds tested showed mutagenic activity in at least one system. It should be stressed, however, that this represents a group of compounds preselected as likely to be mutagenic. We have little doubt that a more extensive search among the literature already in the files of Environmental Mutagen Information Center (EMIC) of the Environmental Mutagen Society would probably reveal that many more of the commonly used pesticides possess mutagenic activity.

It is apparent from the literature that there has been

no large-scale testing of pesticides for mutagenic activity. Existing reports are therefore sporadic. Most of the tests for mutagenic activity of pesticides have been done on plants. The reason for this is that techniques for the accurate scoring of chromosome aberrations in plants were available many years before similar techniques were available for mammalian material. However, a good correlation has been demonstrated for chromosome-breaking agents tested in plant and mammalian cell systems.

A detailed summary table of literature reviewed is given in the Appendix. The following examples are presented for illustration only.

**Fumigants** These are generally used in confined areas and will reach the human population only if they are persistent. Since fumigants are often highly reactive, it is possible that their breakdown products, rather than the original fumigants, will reach the population. For example, ethyleneoxide is a highly reactive alkylating agent which has been shown to be mutagenic in many systems. If chlorine ions are present in fumigated material, epichlorhydrin will be formed. This compound, stable enough to persist in marketed material, induces point mutations in *Neurospora crassa.* Several other fumigants which have shown mutagenic action include formaldehyde, active in *Neurospora* and *Drosophila*, and ethylene dibromide, active in *Neurospora.*

**Mercury pesticides** Twelve of the 317 pesticides in a recent listing contain mercury. Eight hundred thousand pounds of mercury, representing 15 percent of the total amount of mercury marketed per annum in the United States, are used in the production of pesticides. Most of the remaining 85 percent are used in mildew-resistant paints. In nature, most mercurial compounds are finally converted into methyl-mercury, which then accumulates in fish and shellfish. Human consumption of such seafood may lead to accumulation of methyl-mercury to even lethal levels. Methyl-mercury causes nondisjunction in *Drosophila*. High levels of chromosome aberrations have been found among heavy fish eaters in Sweden.[2] In Japan, deaths and teratogenic effects have been directly attributed to the intake of mercury-containing seafoods.[4] It is known that mercurial residues can persist up to one hundred years in polluted lakes. The use of many mercury pesticides is now prohibited in Sweden.

**Organophosphate insecticides** Many are triesters of phosphoric acid, and, as such, might be alkylating. The simplest triester is trimethylphosphate: while not a pesticide, it can, however, induce point mutations in *Neurospora crassa* and it is highly active in the dominant lethal mouse test.[1] Dichlorvos, an organophosphate insecticide, is mutagenic in submammalian systems[3, 6]; it is possible that others will be found to be mutagenic when tested.

**Miscellaneous** Captan is an alkylating agent which induces chromosome rearrangements in rats and point mutations in *Neurospora crassa.* Maleic hydrazide breaks chromosomes in many plant systems, although it is inactive in the dominant lethal test in the mouse. Lindane is also similarly active in many plant systems.

Although at present only limited information on the mutagenic action of commonly used pesticides exists in the literature, it is likely that automated procedures will be required to keep up with an expanding literature.

References


1. Epstein, S. S., Bass, W., Arnold, E., and Bishop, Y. 1970. The mutagenicity of trimethyl phosphate in mice. *Science 168:* 584–586.
2. Lofroth, G. Personal communication.
3. Michalek, S. M., and Brockman, H. E. 1969. A test of mutagenicity of Shell "No-Pest" strip insecticide in *Neurospora crassa. Neurospora Newsletter 14:* 8.
4. Miller, R. W. 1967. Prenatal origins of mental retardation: Epidemiological approach. *J. Pediatrics 71:* 455–458.
5. Neumeyer, J., Gibbons, D., and Trask, H. 1969. Pesticides. *Chemical Week,* Apr. 12, pp. 38–68, and Apr. 26, pp. 38–68.
6. Sax, K., and Sax, H. J. 1968. Possible mutagenic hazards of some food additives, beverages, and insecticides. *Japan, J. Genet. 43:* 89–94.

# 8 Summary and Recommendations

Among the many new chemicals in our increasingly complex environment, a number are already known to be mutagenic, i.e., capable of producing genetic damage. When genetic damage occurs, the burden of hereditary defects in future generations is increased. One potential genetic hazard comes from pesticides. Although we can point to no pesticide now in wide use that has been demonstrated to be mutagenic, the overwhelming majority have not been adequately tested, although appropriate methodologies are now available.

We define a mutation as any inherited alteration in genetic material. Such alterations, in exposed individuals, may lead to cancer and to teratogenic effects. Our main concern, however, is for the descendants of these individuals; for such changes lead to a wide range of abnormalities, mental retardation, physical and mental disease, and all the other inherited weaknesses and debilities to which man is susceptible. Since these effects will occur in future generations and may be apparent only many generations removed, by the time the effect is noticed, the damage is already irreversible. It is, therefore, urgent that any mutagenic chemicals to which the population is exposed be promptly identified.

There are now about 400 substances that, in various forms and combinations, are currently used as pesticides. It is feasible to test all of these in the near future for mutagenicity in systems that are simple and precise, and yet relevant to man.

For these and other reasons detailed in the report, we make the following recommendations: (a) All currently used pesticides should be tested in the near future in four systems (see chapter 4). Pesticides should be tested at concentrations substantially higher than those to which the human population is likely to be exposed. Test procedures should reflect routes of human exposure. Apart from the obvious route of ingestion, particular attention should be directed to inhalation, especially for pesticide aerosols and for vaporizing pesticide strips, which are widely used domestically. (b) Pesticides found to be inactive in all these tests may be regarded as safe, unless other evidence of mutagenicity appears. Use of mutagenic pesticides must be rigorously restricted, unless thorough and disinterested study demonstrates convincingly that the benefit outweighs the risk. (c) No new pesticide should be registered until tested for mutagenicity. (d) Some impartial scientific group or commission, representing a wide range of agency and nonagency interests, should be charged with responsibility for continued surveillance of the whole problem of pesticide toxicology in general and mutagenesis in particular.

# Appendixes

## 1 Tabulations and Cross Index of Pesticides

The tabulations are based on Neumeyer, J., Gibbons, D., and Trask, H. 1969. Pesticides. *Chemical Week*, Apr. 12, pp. 43–66, and April 26, pp. 41–68.

The pesticides are listed alphabetically by their common names, established by the K–62 Committee of the U.S.A. Standards Institute (formerly ASA) or by names established by the former interdepartmental Committee on Pest Control. Alternative names are also listed; these include common names assigned by the Entomological Society of America and the Weed Society of America, and trade and proprietary names. Chemical names are those used by the Chemical Abstracts Service of the American Chemical Society; they are shown with the structural formulas. All names are cross-referenced in an index at the end of the tables. Major-use listings are based on general knowledge, advice supplied by manufacturers, or reference texts. The cross index is a guide to the tabulations. Common names are the basis for alphabetization and are cited in **boldface** type. Alternative common names and chemical names are in lightface, followed by the main name in boldface. Entries marked with an asterisk are registered trademark names.

| Pesticides common name and synonyms | Chemical name and formula | Major uses | Manufacturers |
|---|---|---|---|
| **Abate*** | 0,0,0′,0′-Tetramethyl 0,0′-thiodi-*p*-phenylene phosphorothioate | Mosquito and midge larvicide | Cyanamid |
| **Acaralate***<br>Chloropropylate* | Isopropyl 4,4′-dichlorobenzilate | Mite control on apples and pears | Geigy |
| **Acrolein**<br>Aqualin*<br>acrylaldehyde | 2-Propenal<br>$CH_2{=}CH{-}CHO$ | Control of submersed and floating aquatic vegetation and algae | Shell |
| **Acrylonitrile**<br>Acritet* | Acrylonitrile<br>$CH_2{=}CHCN$ | Fumigant for stored grain, tobacco, nuts, and dates to control insects | Stauffer<br>Carbide<br>Monsanto<br>Cyanamid |
| **Akton***<br>SD9098 | 0-[2-chloro-1-(2,5-dichlorophenyl)-vinyl]0,0-diethyl phosphorothioate | Soil insecticide—control of lawn chinch bugs and sod webworms in St. Augustine turf | Shell |

| Pesticides common name and synonyms | Chemical name and formula | Major uses | Manufacturers |
|---|---|---|---|
| **Aldrin**<br>Drinox*<br>Seedrin | 1,2,3,4,10,10-Hexachloro-1,4,4a,5,8,8a-hexahydro-1,4 endo-exo-5,8-dimethano-naphthalene (not less than 95% in Aldrin) 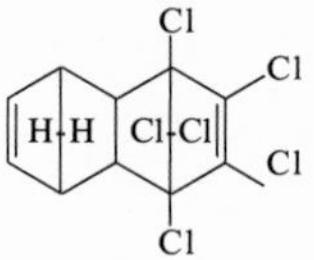 | Broad-spectrum insecticide used for the control of soil insects, cotton insects, turf pests, grasshoppers, corn rootworms, wire worms, white grubs, etc. | Shell |
| **Allethrin**<br>Allyl homolog of cinerin I | DL-2-Allyl-4-hydroxy-3-methyl-2-cyclopenten-1-one esterified with a mixture of *cis* and *trans* DL-chrysanthemum monocarboxylic acid 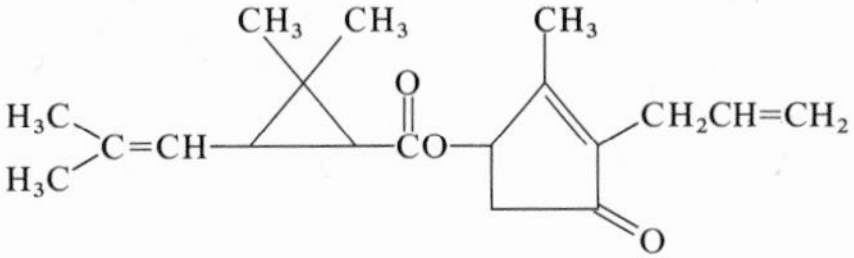 | Control of housefly in household space sprays and aerosols; postharvest application in fruits, vegetables and mushrooms for control of fruit fly; used with synergist | MGK |
| **Allyl alcohol** | 2-Propene-1-ol<br>$CH_2{=}CH{-}CH_2OH$ | Control of weed seeds and certain damping-off fungi; contact herbicide for control of weed and grass seed and applied as a drench to tobacco seed beds | Shell<br>Dow |
| **Aluminum phosphide**<br>Phostoxin*<br>Delicia | Aluminum phosphide<br>$Al_3P_2$ | Fumigant to control stored products insects in silos and warehouses; reacts with water to release phosphine gas | Hollywood |

| | | | |
|---|---|---|---|
| **Ametryne*** | 2-(Ethylamino)-4-(isopropylamino) = 6-(methylthio)-s-triazine 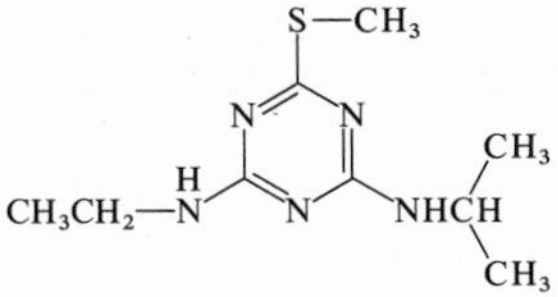 | For weed control in pineapple and sugarcane | Geigy |
| **Amiben***<br>Amoben<br>Chloramben<br>Verbigen | 3-Amino-2,5-dichlorobenzoic acid 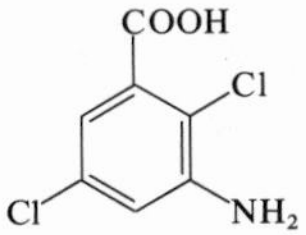 | Selective pre-emergence herbicide for soybeans, peanuts, sweet potatoes, Irish potatoes, and cucurbits | Amchem |
| **Amitrole**<br>Weedazol*<br>Cytrol* | 3-Amino-1,2,4-triazole 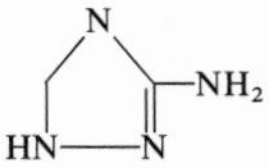 | For control of annual grasses and broadleaf weeds, perennial broadleaf weeds and grasses, cattails, poison ivy, and certain aquatic weeds in marshes and drainage ditches; for nonfood products only | Cyanamid<br>Amchem |
| **AMS**<br>Ammate* | Ammonium sulfamate 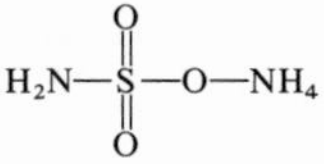<br>$H_2N-SO_2-O-NH_4$ | Contact and translocated herbicide for brush and poison ivy and other undesirable woody plants | DuPont |
| **Anthraquinone**<br>Morkit* | 9,10-Anthraquinone 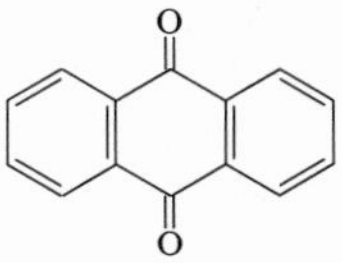 | Bird repellent; treatment of seeds of pines, cereals, vegetables, legumes | Ciba<br>Winthrop |

| Pesticides common name and synonyms | Chemical name and formula | Major uses | Manufacturers |
|---|---|---|---|
| **Antiresistant/DDT**<br>Pramex | N,N-di-n-butyl-*p*-chlorobenzenesulfonamide<br>$Cl-C_6H_4-SO_2-N(C_4H_9)_2$ | Restores DDT activity on resistant strains—synergist | Penick<br>WARF |
| **Antu*** | α-Napthylthiourea<br>$C_{10}H_7-NH-C(=S)-NH_2$ | Toxic to rodents, particularly Norway rat | Penick |
| **Aramite***<br>Niagaramite*<br>Aracide* | 2-(*p*-tert-butylphenoxy)-1-methylethyl 2-chloroethyl sulfite<br>$ClCH_2CH_2\text{-}O\text{-}S(=O)\text{-}O\text{-}CH(CH_3)\text{-}CH_2\text{-}O-C_6H_4-C(CH_3)_3$ | Adulticide for ornamentals, cotton and nonbearing fruit trees; toxic to phytophagous mites | Uniroyal |
| **Arsenic acid** | Arsenic acid<br>$HO-As(OH)_2{=}O\cdot\frac{1}{2}H_2O$ | Cotton desiccant | Pennsalt<br>FMC |

| | | | |
|---|---|---|---|
| **Aspon***<br>NPD* | 0,0,0,0-Tetrapropyl dithiopyrophosphate 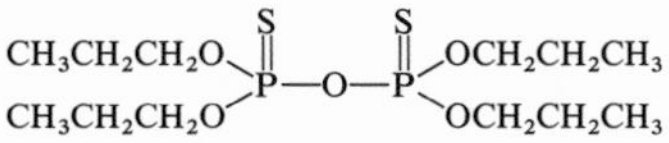<br>$CH_3CH_2CH_2O$ S S $OCH_2CH_2CH_3$<br>P—O—P<br>$CH_3CH_2CH_2O$ $OCH_2CH_2CH_3$ | Highly effective for control of chinch bugs in turf | Stauffer |
| **Atratrone**<br>Gesatamin<br>G-32293 | 2-(Ethylamino)-4-(isopropylamino) 6-methoxy-s-triazine 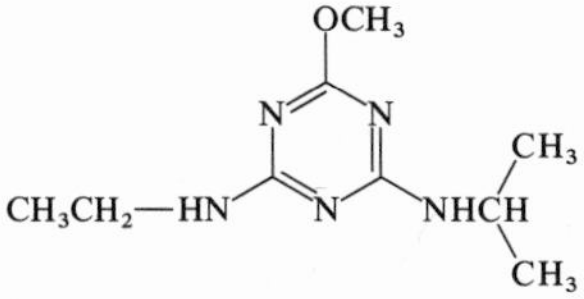<br>$OCH_3$<br>N N<br>$CH_3$<br>$CH_3CH_2$—HN N NHCH<br>$CH_3$ | Experimental herbicide; absorbed by both leaves and roots, unlike simazine, which is taken up only by roots | Geigy |
| **Atrazine**<br>Aatrex*<br>Atratol*<br>Rack Granular*<br>Gesaprim*<br>Primatol A* | 2-Chloro-4-ethylamino-6-isopropylamino-s-triazine 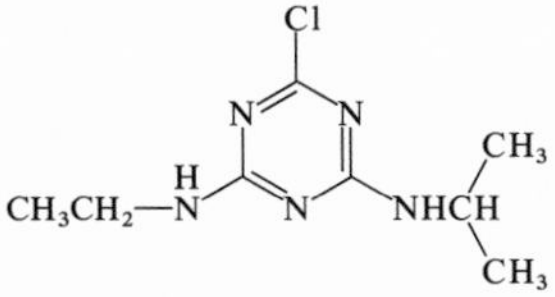<br>Cl<br>N N<br>$CH_3$<br>H<br>$CH_3CH_2$—N N NHCH<br>$CH_3$ | For season-long weed control in corn and sorghum, for weed control in certain other crops, in noncrop areas and industrial sites; selective pre- and postemergence herbicide | Geigy |
| **Azinphosethyl**<br>Ethyl Guthion* | 0,0-diethyl-S-[4-oxo-1,2,3-benzotriazin-3(4H)-ylmethyl]-phosphorodithioate 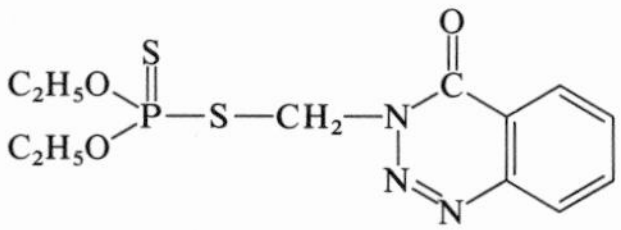<br>O<br>S C<br>$C_2H_5O$<br>P—S—$CH_2$—N<br>$C_2H_5O$<br>N<br>N | Insecticide for cotton and potatoes; not registered for any use on crops in the U.S.; widely used on vegetables and fruits in Europe | Chemagro |

| Pesticides common name and synonyms | Chemical name and formula | Major uses | Manufacturers |
|---|---|---|---|
| **Azinphosmethyl**<br>Guthion*<br>Gusathion* | 0,0-Dimethyl-S-[4-oxo-1,2,3-benzotriazin-3(4H)-ylmethyl]-phosphorodithioate 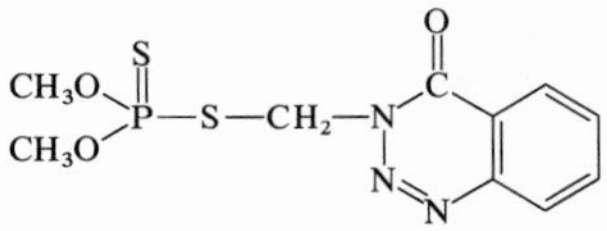 | Broad spectrum insecticide for cotton, tobacco, deciduous fruits, strawberries, beans, cole crops, potatoes, tomatoes, other fruits, field and vegetable crops; ornamentals, flowers, shrubs and trees | Chemagro |
| **Azobenzene**<br>Diphenyl diimide<br>Azofume<br>Benzofume<br>Azobenzide | Azobenzene 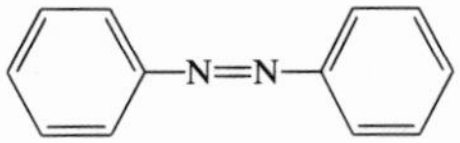 | Greenhouse fumigant for active and egg mite states; not for food crops | Eastern |
| **Azodrin***<br>SD9129 | Dimethyl phosphate of 3-hydroxy-N-methyl-cis-crotonamide 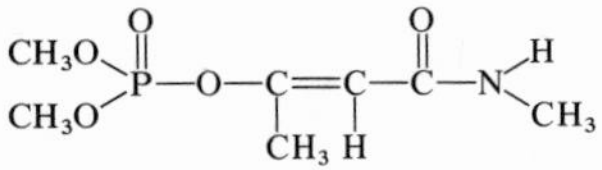 | Controls pests such as certain economically important insects that attack cotton plants | Shell |
| **Bacillus thuringiensis**<br>Agritol*<br>Bakthane*<br>Larvatrol*<br>Thuricide*<br>Biotrol BTB* | Microbial insecticide containing viable spores of Bacillus thuringiensis on dust or stabilized suspension form | Specific for larval stages of lepidopterous insects; stomach poison to insect | Rohm & Hass<br>IMC<br>Stauffer<br>Thompson-Hayward<br>Pennsalt |
| **Bandane*** | Polychlorobicyclopentadiene isomers 60–62% chlorine | Pre-emergence herbicide for crabgrass; low phytotoxicity to turf grasses; also controls turf insects such as ants and grubs | Velsicol |

| | | | |
|---|---|---|---|
| **Barban**<br>Carbyne* | 4-Chloro-2-butynyl-*m*-chlorocarbanilate 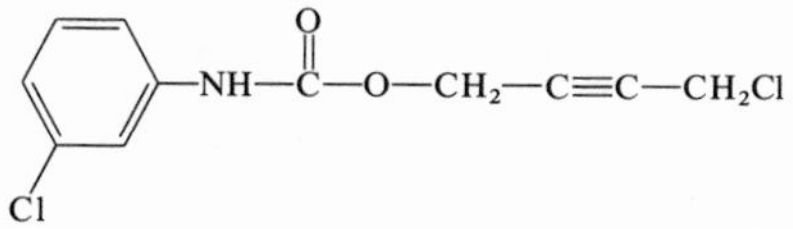 | Control of wild oats in barley, flax, lentils, mustard for oil, peas, safflower, sugar beets and wheat | Gulf |
| **Barthrin**<br>6-chloropiperonyl chrysanthemumate | 6-Chloropiperonyl 2,2-dimethyl-3-(2-methylpropenyl)cyclopropane-carboxylate 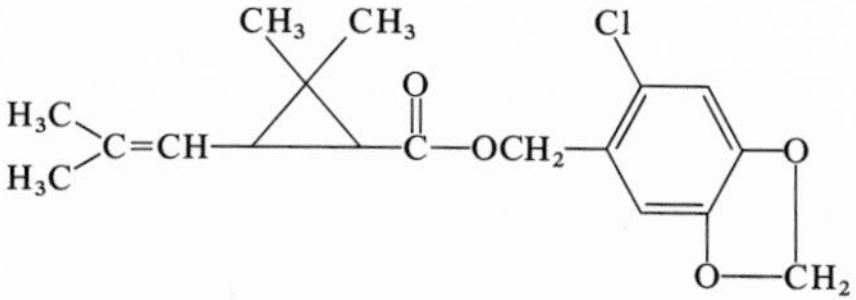 | Household space sprays and aerosols | MGK |
| **Baygon***<br>Bay* 39007<br>Propoxur<br>Unden* | 2-Isopropoxyphenyl N-methylcarbamate 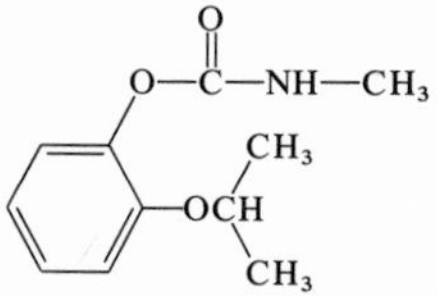 | Mosquitos, flies, ants, cockroaches, spiders, chinch bugs, sod-webworms and sand flies; rapid knockdown and good plant compatibility; systemic activity against thrips, aphids, leaf miners, leafhoppers, and scales | Chemagro |
| **Bayluscide***<br>Chlonitralid<br>Yomesan | 2′,5-Dichloro-4′-nitrosalicylanilide ethanolamine 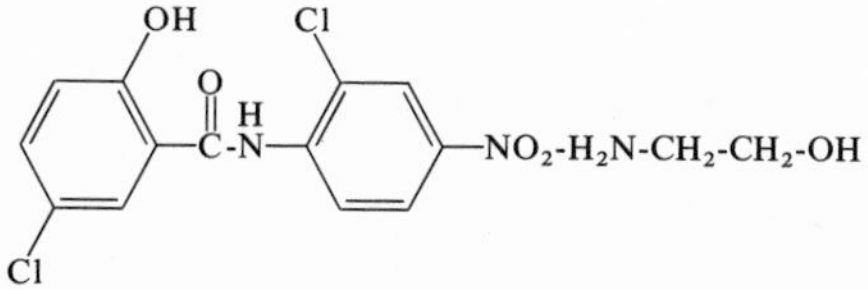 | Registered for use in flowing streams and impounded water for the control of fresh-water snails and sea lamprey | Chemagro |

| Pesticides common name and synonyms | Chemical name and formula | Major uses | Manufacturers |
|---|---|---|---|
| **Benefin**<br>Balan*<br>Binnell* | N-Butyl-N-ethyl-α,α,α-trifluoro-2,6-dinitro-*p*-toluidine 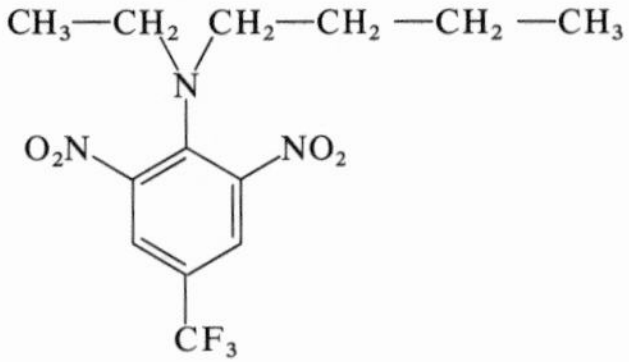<br>$CH_3—CH_2$ $CH_2—CH_2—CH_2—CH_3$ N; $O_2N$, $NO_2$, $CF_3$ | Pre-emergence herbicide, for control of annual grasses and broadleaf weeds in peanuts; direct seeded lettuce; established turf, alfalfa, red clover, birdsfoot trefoil, and sweet clover grown for seed | Elanco |
| **Benlate*** | Methyl 1-(butylcarbamoyl)-2 benzimidazole carbamate 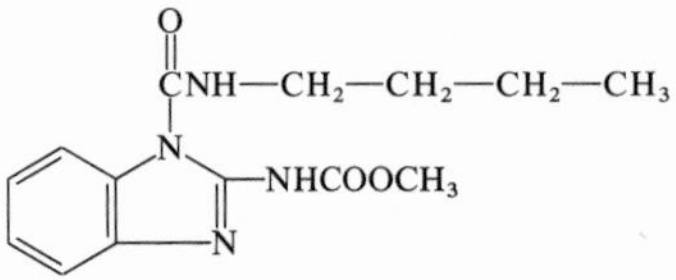<br>O ‖ $CNH—CH_2—CH_2—CH_2—CH_3$; N; $NHCOOCH_3$; N | Expected to be available in '69 for nonfood use (turf and ornamentals); fungus control for fruit, vegetables and field crops | DuPont |
| **Bensulide**<br>Prefar*<br>Betasan*<br>R-4461 | S-(0,0-Diisopropyl phosphorodithioate) ester of N-(2-mercaptoethyl)benzene-sulfonamide 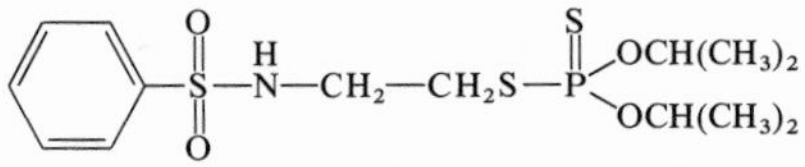<br>$S(=O)_2—NH—CH_2—CH_2S—P(=S)(OCH(CH_3)_2)_2$ | Selective pre-emergence herbicide effective control of annual grasses, and broadleaf weeds in dichondra and grass lawns, especially effective for seasonal control of crabgrass and annual bluegrass in grass and dichondra lawns | Stauffer |

| | | | |
|---|---|---|---|
| **Betanal***<br>Phenmedipham | Methyl *m*-hydroxycarbanilate *m*-methylcarbanilate 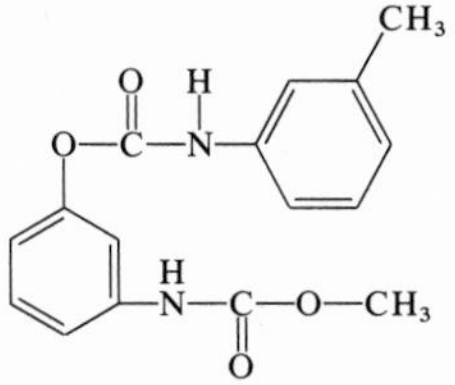 | Selective herbicide for use on sugar beets, red beets | Morton |
| **BHC**<br>Benzene hexachloride<br>Benzahex*<br>Benzex*<br>Gammexane*<br>Isotox*<br>Lintox* | Several isomers of benzenehexachloride containing 12–14% of gamma isomer 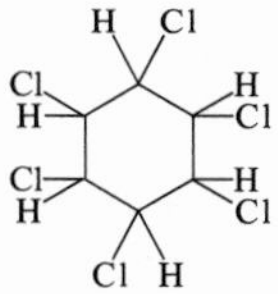 | Broad-spectrum insecticide for fruits, legumes, cole crops, cucurbits, tomatoes, other vegetables | Diamond<br>Hooker |
| **Bidrin***<br>SD3562 | Dimethyl phosphate of 3-hydroxy-N,N-dimethyl-*cis*-crotonamide 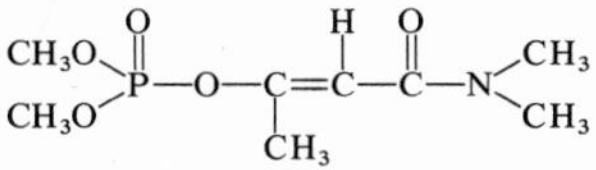 | Certain economically important insects and mites that attack ornamentals and principal field, forage, vegetable, and fruit crops and cotton | Shell |

| Pesticides common name and synonyms | Chemical name and formula | Major uses | Manufacturers |
|---|---|---|---|
| **Binapacryl**<br>Morocide* | 2-sec-butyl-4,6-dinitrophenyl-3-methyl-2-butenoate 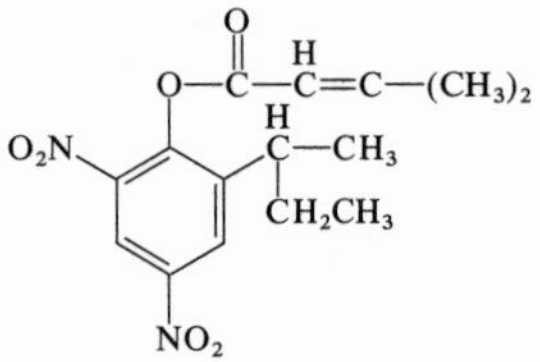 | Contact poison for spider mites and powdery mildew on fruit and nut trees | FMC |
| **Biphenyl**<br>Diphenyl | Biphenyl 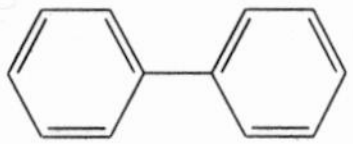 | Impregnating fruit wraps | Dow |
| **Bomyl*** | Dimethyl 3-hydroxyglutaconate dimethyl phosphate 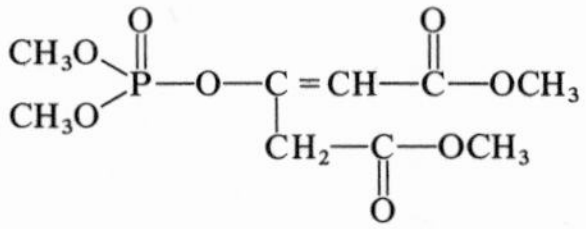 | Effective as a contact-residual treatment for a wide variety of insects and mites and houseflies | Allied |
| **Borax***<br>Trona<br>Borascu*<br>Polybor3*<br>Boro-Spray*<br>Neobor*<br>Tronabor* | Sodium tetraborate decahydrate<br>$Na_2B_4O_7.10H_2O$ | For control of a wide variety of weeds; has limited use in ant poisons and for fly control around refuse and manure piles; as a larvacide, which includes control of the common housefly and also of dog hook-worm and swine kidney worm; nonfood use, excluding poultry houses | American Potash<br>U.S. Borax |

| | | | |
|---|---|---|---|
| **Bordeaux mixture** | Mixture of copper sulfate and hydrated lime; precise structure unknown | Foliage fungicide for potatoes, tomatoes, cucurbits, bananas, stone and pome fruits | (Tank mixed on farm) |
| **Bromoxynil**<br>Buctril*<br>Brominal* | 3,5-Dibromo-4-hydroxybenzonitrile 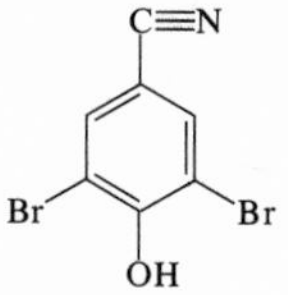<br>$C{\equiv}N$; Br; Br; OH | Selective herbicide for weeds and in seedling cereal grains | Amchem<br>Chipman<br>May & Baker |
| **Butonate** | Dimethyl (2,2,2-trichloro-1-hydroxyethyl) phosphorate ester of butyric acid 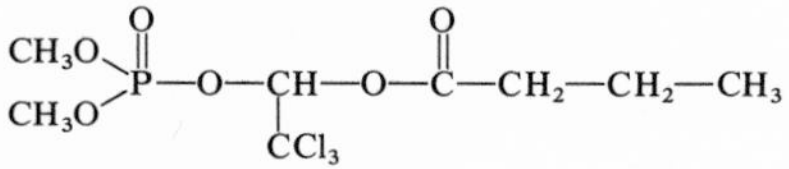<br>$(CH_3O)_2P(=O)-O-CH(CCl_3)-O-C(=O)-CH_2-CH_2-CH_3$ | Control of flies, cockroaches, ants, and other household insects | Prentiss |
| **Butoxypolypropylene glycol**<br>Stabilene*<br>Crag* Fly Repellent | Butoxypolypropylene glycol 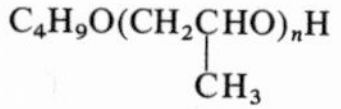<br>$C_4H_9O(CH_2CH(CH_3)O)_nH$ | Repellent toward biting insects in dairy and livestock insecticide formulations | Carbide |
| **Butylate**<br>Sutan* | S-Ethyl diisobutylthiocarbamate 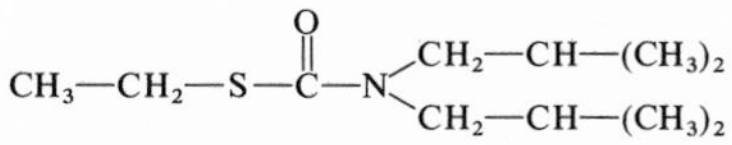<br>$CH_3-CH_2-S-C(=O)-N(CH_2-CH-(CH_3)_2)_2$ | Selective preplant herbicide for control of annual grasses, some broadleaf weeds, and nutgrass in corn | Stauffer |
| **Butyl mesityl oxide oxalate**<br>Indalone*<br>Butopyronoxyl | Butyl 3,4-dihydro-2,2-dimethyl-4-oxo-1-2H-pyran-6-carboxylate 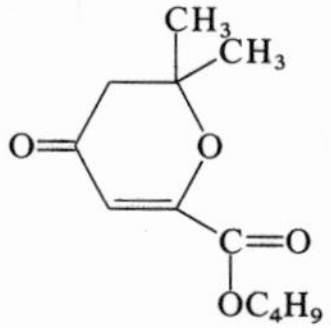<br>$CH_3$; $CH_3$; O=; O; C=O; $OC_4H_9$ | Repellent to biting insects; no insecticidal activity | U.S.P.<br>FMC |

| Pesticides common name and synonyms | Chemical name and formula | Major uses | Manufacturers |
|---|---|---|---|
| **Bux Ten***<br>Bux<br>Ortho 5353 | Mixture of *m*(1-methylbutyl)phenylmethyl carbamate (75%) and *m*(1-ethylpropyl) phenyl methylcarbamate (25%) 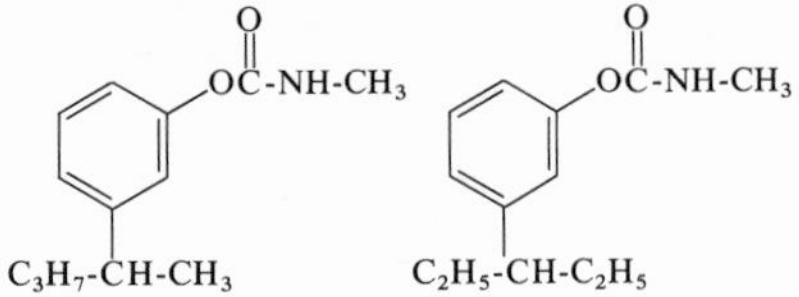<br>(structures: m-substituted phenyl OC(=O)-NH-$CH_3$ with $C_3H_7$-CH-$CH_3$; and with $C_2H_5$-CH-$C_2H_5$) | Effective against larvae of corn rootworm, also against resistant western corn rootworm | Chevron |
| **Cacodylic acid**<br>Dimethylarsinic acid<br>Phytar* 560<br>Rad-E-Cate 35*<br>Arsan | Dimethyl arsinic acid 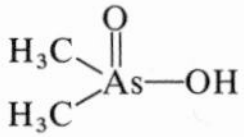<br>$(H_3C)_2As(=O)$—OH | General weed control; cotton defoliant | Vineland<br>Ansul |
| **Cadmium-calcium-copper-zinc chromate complex**<br>Miller 531*<br>Crag Turf Fungicide 531* | Mixture of cadmium-calcium-copper-zinc sulfate-chromate<br>$6CdO, 10CaO, 25CuO, 10ZnO, 25SO_3, 10CrO_3, 17OH_2O$ | Turf disease control | Miller<br>Carbide |
| **Calcium arsenate** | Calcium arsenate<br>$[Ca_3(As_3O_4)_2]\cdot Ca(OH)_2$<br>Precise structure unknown | Vegetables and fruit as bait | Chevron<br>Chipman<br>FMC<br>Pennsalt |
| **Calcium cyanamide**<br>Aero | Calcium cyanamide<br>$Ca{=}N{-}C{\equiv}N$ | Cotton defoliation in humid East; control of certain soil-borne diseases | Cyanamid |

| | | | |
|---|---|---|---|
| **Calcium cyanide**<br>Cyanogas* | Calcium cyanide 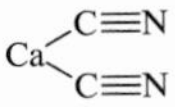 | Control of insect and rodent pests in flour mills, seed houses, grain elevators; in city and state public health programs for rat control; tent fumigation of citrus trees; fumigation of greenhouses and mushroom houses | Cyanamid |
| **Captan***<br>Orthocide* | N-Trichloromethylthio-4-cyclohexene 1,2-dicarboximide 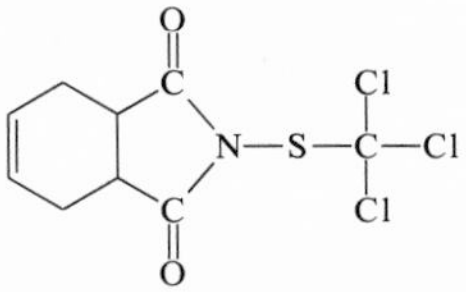 | As a protectant eradicant fungicide for fruits, vegetables, and flowers in control of scabs, blotches, rots, mildew, etc. | Stauffer<br>Chevron |
| **Carbaryl**<br>Sevin* | 1-Naphthyl N-methylcarbamate 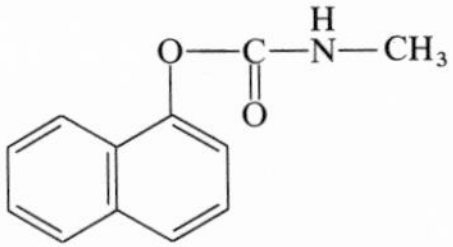 | Control of insects on fruits, vegetables, forage, cotton, and other economic crops, as well as poultry and pets | Carbide |
| **Carbofuran**<br>Furadan* | 2,3-Dihydro-2,2-dimethyl-7-benzofuranyl methylcarbamate 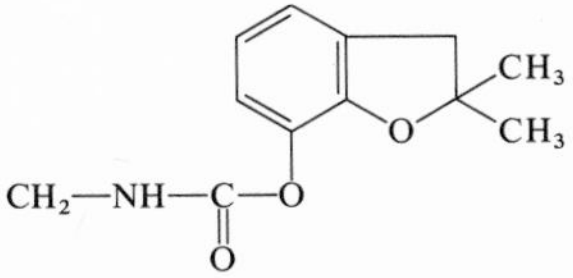 | Systemic and contact insecticide and nematocide; promising for corn rootworm, alfalfa weevil; broad-spectrum insecticide for use on tobacco, rice, peanuts, sugar-cane, fruits and vegetables | FMC |

| Pesticides common name and synonyms | Chemical name and formula | Major uses | Manufacturers |
|---|---|---|---|
| **Carbon disulfide**<br>Carbon bisulfide | Carbon disulfide<br>$S{=}C{=}S$ | Fumigant for stored grain and commodities; soil fumigation for fungi and deep-rooted perennials | Stauffer<br>FMC<br>PPG |
| **Carbon tetrachloride** | Carbon tetrachloride<br>$CCl_4$ | Stored grain and industrial fumigant; also in combination with flammable fumigants | Dow<br>Stauffer<br>Diamond<br>FMC |
| **Carbophenothion**<br>Trithion* | S-[(*p*-Chlorophenylthio)methyl] 0,0-diethyl phosphorodithioate<br>$(C_2H_5O)_2P(=S){-}S{-}CH_2{-}S{-}C_6H_4{-}Cl$ | Useful on wide range of fruit and nut, vegetable and fiber crops; particularly effective as a miticide | Stauffer |
| **Carzol***<br>Formetanate hydrochloride<br>EP-332<br>ENT-27566 | *m*[[(Dimethylamino) methylene]-amino] phenyl methylcarbamate hydrochloride<br>$(H_3C)_2N{-}CH{=}N{-}C_6H_4{-}O{-}C(=O){-}NH{-}CH_3$ | Effective for control of spider mites, rust mites, aphids, thrips, leafhoppers, slugs and snails on horticultural and ornamental plants | Morton |
| **CDAA**<br>Randox*<br>Limit* | N,N-Diallyl-2-chloroacetamide<br>$ClCH_2{-}C(=O){-}N(CH_2{-}CH{=}CH_2)_2$ | For pre-emergence application to control weedy annual grasses in corn, soybeans, sorghum, onions, cabbage, sweet potatoes, tomatoes, celery, sugarcane and Irish potatoes | Monsanto |

| | | | |
|---|---|---|---|
| **CDEC**<br>Vegadex* | 2-Chloroallyl diethyldithiocarbamate 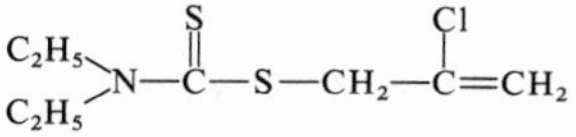<br>$(C_2H_5)_2N{-}C({=}S){-}S{-}CH_2{-}C(Cl){=}CH_2$ | Pre-emergence selective herbicide; controls many annual grasses and broadleaf weeds in vegetable crops, including leaf and cole crops, corn, beans, salad crops, tomatoes, cantaloupes, ornamentals, and shrubbery | Monsanto |
| **Ceresan L***<br>Granosan L | Mixture of methylmercury 2,3-dihydroxypropyl mercaptide and methylmercury acetate 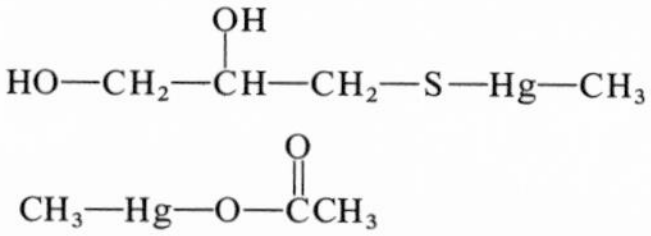<br>$HO{-}CH_2{-}CH(OH){-}CH_2{-}S{-}Hg{-}CH_3$<br>$CH_3{-}Hg{-}O{-}C({=}O)CH_3$ | To control certain seed-borne diseases and to reduce losses from seed decay and seedling blights of wheat, rye, barley, oats, sorghum and millet, safflower, flax, rice and cotton; in spray applications for control of snow mold of wheat; as a soak treatment to prevent basal rot of gladiolus bulbs | DuPont |
| **Ceresan M***<br>Ceresan M-DB* | N-(Ethylmercury)-*p*-toluenesulfonanilide 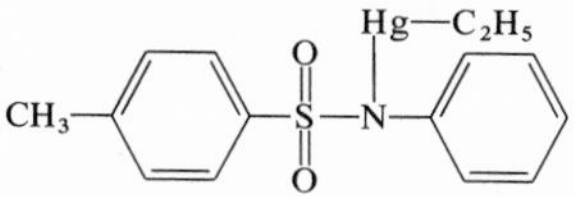<br>$CH_3{-}C_6H_4{-}SO_2{-}N(Hg{-}C_2H_5){-}C_6H_5$ | For control of certain seed-borne diseases and to reduce losses from seed decay and damping-off | DuPont |
| **Chloranil**<br>Spergon | Tetrachloro-*p*-benzoquinone 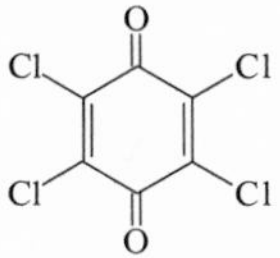<br>$C_6Cl_4O_2$ (structure) | Seed dressing for peas, beans, and vegetables; foliage spray and dust for downy mildew of cabbage and related crops | Uniroyal |
| **Chlorazine** | 2-Chloro-4,6-bis(diethylamino)-s-triazine 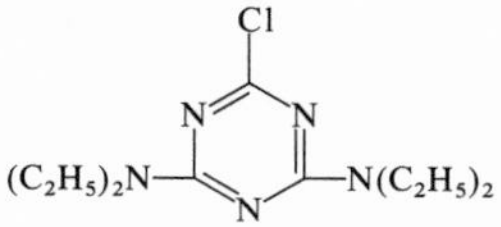<br>Cl on triazine ring; $(C_2H_5)_2N{-}$ and ${-}N(C_2H_5)_2$ substituents | | Geigy |

| Pesticides common name and synonyms | Chemical name and formula | Major uses | Manufacturers |
|---|---|---|---|
| **Chlordane**<br>Chlor Kil*<br>Corodane*<br>Ortho-Klor*<br>Synklor* | Mixture of 60% 1,2,3,4,5,6,7,8,8-Octa-chloro-3a,4,7,7a-tetrahydro-4,7-methanoindane and 40% related compounds 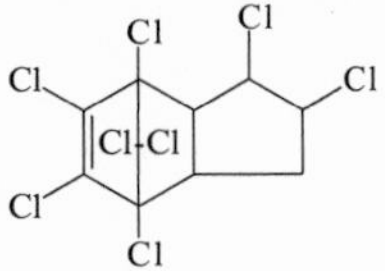 | Broad-spectrum insecticide for household or institutional pest control; lawn termite control; for vegetable and field crops | Velsicol |
| **Chlorobenzilate**<br>Acaraben* | Ethyl 4,4'-dichlorobenzilate 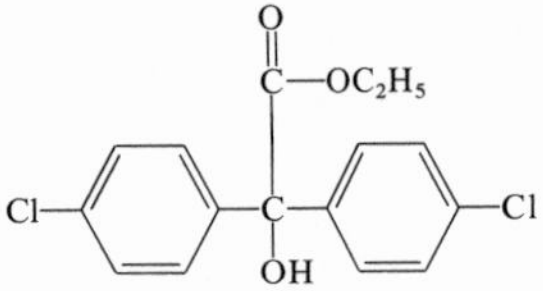 | Control of mites on certain agricultural crops and ornamentals | Geigy |
| **Chloroform** | Chloroform 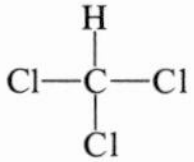<br>$CHCl_3$ | Insecticide; grain fumigant mixture contains 73.2% $CHCl_3$, 26.8% $CS_2$; screw worm control on animals | |
| **Chloroneb**<br>Demosan* | 1,4-Dichloro-2,5-dimethoxybenzene 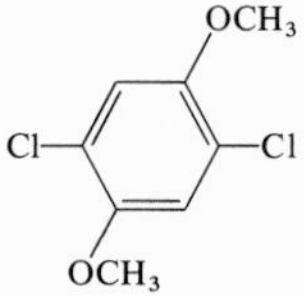 | As a supplemental seed treatment for cotton, soybeans (seed crop only) and beans (seed crop only) or as an in-furrow soil treatment at planting time, for the control of seedling diseases such as seedling blights, pre- and postemergence damp-off | DuPont |

| Name | Chemical name | Use | Manufacturer |
|---|---|---|---|
| **Chloropicrin**<br>Larvacide*<br>Picfume*<br>Acquinite*<br>Nemax* | Trichloronitromethane<br>$Cl-C(Cl)(Cl)-NO_2$ | Fumigation of soil before planting; controls nematodes, insects, fungi, bacteria | Dow<br>Morton<br>Niklor |
| **Chloroxuron**<br>Tenoran*<br>C-1983 | 3-[*p*-(*p*-chlorophenoxy)phenyl]-1,1-dimethylurea<br>$(H_3C)_2N-C(=O)-N(H)-C_6H_4-O-C_6H_4-Cl$ | Weed control in soybeans, strawberries, carrots, onions, and other crops | Ciba |
| **Chlorphenamidine**<br>Galecron*<br>Fundal | N,N-Dimethyl-N′-(2-methyl-4-chlorophenyl)-formamidine<br>$Cl-C_6H_3(CH_3)-N=C(H)-N(CH_3)_2$ | Adulticide and as ovicide against all species of mites, including resistant and susceptible strains | Ciba<br>Morton |
| **Chlorpropham**<br>Chloro IPC<br>CIPC | Isopropyl *m*-chlorocarbanilate<br>$Cl-C_6H_4-N(H)-C(=O)-O-CH(CH_3)_2$ | Highly selective preemergence and early postemergence herbicide; effective control of many annual grassy and broadleaved weeds | PPG |
| **Ciodrin*** | Dimethyl phosphate of α-methylbenzyl 3-hydroxy-cis-crotonate<br>$(CH_3O)_2P(=O)-O-C(CH_3)=C(H)-C(=O)-O-C(H)(CH_3)-C_6H_5$ | Controls flies, lice, and ticks on dairy and beef cattle, swine, goats, and sheep; also in premises | Shell |

| Pesticides common name and synonyms | Chemical name and formula | Major uses | Manufacturers |
|---|---|---|---|
| **CMA**<br>Calcium acid methyl arsenate<br>Calar | Calcium methanearsonate 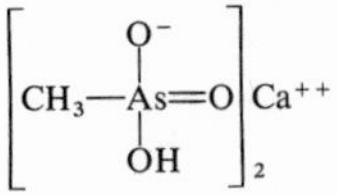<br>$[CH_3-As(=O)(O^-)(OH)]_2 Ca^{++}$ | Crabgrass control | Vineland |
| **Copper carbonate, basic**<br>Malachite green<br>Cupric carbonate | Copper carbonate<br>$Cu(OH)_2 \cdot CuCO_3$ | Seed treatment of cereals | Tennessee<br>Mallinckrodt |
| **Copper naphthenate** | Naphthenic acids, copper salts 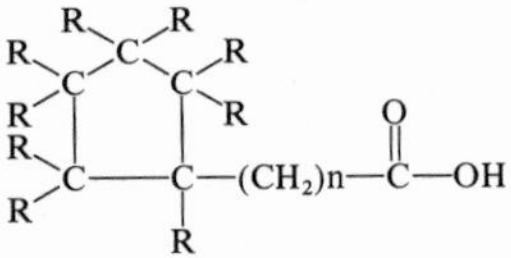<br>$R_2C$–$CR_2$–$CR_2$–$CR_2$–$C(R)$ (ring) $-(CH_2)n-C(=O)-OH$<br>n = 1 to 5, R = H or alkyl | Cotton textiles: wood preservative; field boxes for fruits and vegetables; empty beehives | Harshaw |
| **Copper oleate** | Copper oleate<br>$Cu[O-C(=O)-(CH_2)_7-CH=CH-(CH_2)_7CH_3]_2$ | Fungicide for potatoes, tomatoes and peanuts | Chem. Form.<br>Witco |
| **Copper oxychloride sulfate** | Basic cupric chloride<br>$3Cu(OH)_2 \cdot CuCl_2$ | Foliage fungicide for potatoes, tomatoes, cherries, other vegetables | Harshaw<br>FMC |

| | | | |
|---|---|---|---|
| **Copper-8 -quinolinolate**<br>Bioquin*<br>Copper oxinate | Copper 8-hydroxyquinolinate 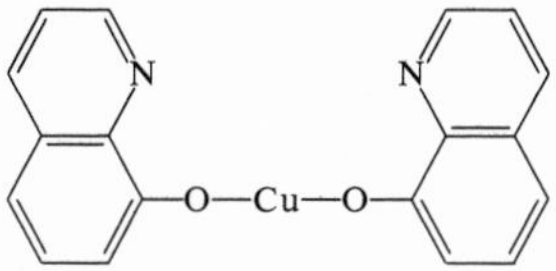 | Packing boxes, food-handling equipment, hampers, harvesting bins | Merck |
| **Copper sulfate** | Copper sulfate<br>$Cu_2SO_4 \cdot 5H_2O$ | Preparation of Bordeaux mixture on farm; manufacture of other copper-containing fungicides | Phelps Dodge<br>Tennessee<br>Calumet<br>Harshaw |
| **Copper zinc chromate**<br>Miller 658<br>Crag Fungicide 658 | Copper-zinc chromate complex<br>$15CuO \cdot 10ZnO \cdot 6CrO_3 \cdot 25H_2O$ | Control of early blight of carrots, *Cercospora* leaf spot of peanuts, bacterial spot on peppers, various diseases of ornamentals | Miller<br>Carbide |
| **Coumaphos**<br>Co-Ral*<br>Asuntol*<br>Muscatox*<br>Resitox* | 0,0-Diethyl 0-3-chloro-4-methyl-2-oxo-2H-1-benzopyran-7-yl-phosphorothioate 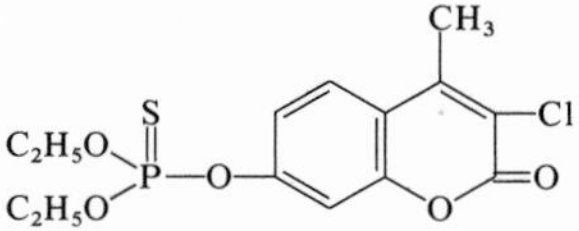 | Control of livestock ectoparasites—e.g., hornflies, screw worms, lice, and ticks; internal cattle grubs | Chemagro |
| **3-CPA**<br>Fruitone* | m-chlorophenoxyacetic acid 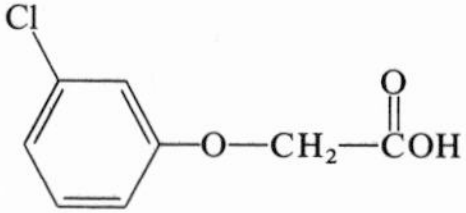 | Plant growth regulator used as peach thinner | Amdal |

| Pesticides common name and synonyms | Chemical name and formula | Major uses | Manufacturers |
|---|---|---|---|
| **4-CPA** | *p*-Chlorophenoxyacetic acid<br>$Cl-C_6H_4-O-CH_2-C(=O)OH$ | Plant growth regulator for bloom set on tomatoes; also used to thin peaches | Dow |
| **Cryolite**<br>Kryocide* | Sodium fluoaluminate<br>$Na_3AlF_6$ | Fruits, vegetables, field crops, and ornamentals | Pennsalt |
| **Cuprous oxide** | Cuprous oxide<br>$Cu_2O$ | Foliage fungicide for cherries, potatoes, tomatoes; seed treatment | Calumet<br>Harshaw<br>Tennessee |
| **Cycloate**<br>Ro-neet*<br>R-2063 | S-Ethyl N-ethyl-N-cyclohexylthio-carbamate<br>$CH_3CH_2-S-C(=O)-N(CH_2-CH_3)-C_6H_{11}$ | Selective herbicide for use on sugar beets, table beets and spinach; effectively controls many annual broad-leafed weeds, annual grasses and nut sedge | Stauffer |
| **Cycloheximide**<br>Actidione* | 3-[2-(3,5-Dimethyl-2-oxocyclohexyl)-2-hydroxyethyl] glutarimide<br>$(CH_3)_2C_6H_7(=O)-CH(OH)-CH_2-C_5H_6O_2N-H$ | Fungicide, effective for the control of powdery mildew on roses and other ornamentals, rusts and leaf spots on lawn grasses, and azalea petal blight | Tuco |

| | | | |
|---|---|---|---|
| **Cycocel*** Chlormequat | 2-Chloroethyltrimethyl ammonium chloride $\left[Cl—CH_2—CH_2—\overset{+}{N}(CH_3)_2—CH_3\right] Cl^-$ | On red poinsettias to produce a more compact plant, also for azaleas for early budded, compact, symmetrical plants; internodes are shortened and stems are thickened; used in Europe to prevent lodging in wheat | Cyanamid |
| **Cyloane*** | Cyclic ethylene (diethoxyphosphinyl) dithiomidocarbonate $(C_2H_5O)_2P(=O)—N=C(SCH_2CH_2S)$ | Systemic insecticide effective against leaf-feeding larvae such as cotton leafworm; also effective against aphids, mites and thrips | Cyanamid |
| **Cypromid** Clobber* S-6000 | 3′,4′-Dichlorocyclopropanecarboxanilide $3,4\text{-}Cl_2C_6H_3—NH—C(=O)—C_3H_5$ | Corn herbicide; controls cocklebur, lamb's-quarters, pigweed, purslane, smartweed, velvet-leaf, wild morning glory, barnyard grass, crabgrass, giant foxtail, green foxtail, Johnson grass, yellow foxtail, and some other broadleafs and grasses in field corn | Gulf |
| **2,4-D** Verton* D DMA-4* WEEDAR* | 2,4-Dichlorophenoxyacetic acid: also used as amine salts and esters $2,4\text{-}Cl_2C_6H_3—O—CH_2—C(=O)—OH$ | Plant growth regulator; post-emergence weed control in cereal grains, corn, pastures and lawns; aquatic weeds | Monsanto Hercules Chipman Thompson-Hayward Diamond Dow |
| **Daconil 2787*** Forturf* | 2,4,5,6-Tetrachlorolsophthalonitrile $C_6Cl_4(C\equiv N)_2$ | Broad-spectrum foliage and fruit protectant fungicide | Diamond |

| Pesticides common name and synonyms | Chemical name and formula | Major uses | Manufacturers |
|---|---|---|---|
| **Dalapon**<br>Dowpon*<br>Radapon* | 2,2 Dichloropropionic acid<br>$CH_3-C(Cl)_2-C(=O)-OH$ | For the control of annual and perennial grasses in sugarcane, sugar beets, corn, potatoes, etc. | Dow |
| **Dasanit***<br>Fensulfothion<br>Terracur P*<br>Bay 2514* | 0,0-Diethyl 0-[*p*-(methylsulfinyl) phenyl— phosphorothioate<br>$(CH_3CH_2O)_2P(=S)-O-C_6H_4-S(=O)-CH_3$ | Nematocide and insecticide for soil insects; some systemic activity; used on field crops, tobacco, vegetables, ornamentals, and turf grasses | Chemagro |
| **Dazomet**<br>Mylone*<br>DMTT<br>Mico-Fume* | Tetrahedro-3,5-dimethyl-2H-1,3,5-thiadiazine-2-thione<br>$CH_3-N$, $N-CH_3$, S, =S (ring) | Soil fungicide, nematocide and weedkiller for use in ornamentals, tobacco, and vegetable seed beds | Stauffer<br>Carbide |
| **2-4-DB**<br>Butyrac<br>Butoxone SB<br>Butyrac ester | 4-(2,4-Dichlorophenoxy)butyric acid: also salt, amine salt and ester formulations<br>$Cl-C_6H_3(Cl)-O-CH_2-CH_2-CH_2-C(=O)-OH$ | Protection of established broadleafs; not effective for grasses; legumes, soybean, and clover crop protection | Amchem<br>Chipman |

| | | | |
|---|---|---|---|
| **DBCP**<br>Fumazone*<br>Dibromochloro-propane<br>Nemagon* | 1,2-Dibromo-3-chloropropane | Soil fumigant and nematocide effective against root-knot nematodes | Shell<br>Dow<br>American Potash |
| **DCNA**<br>Dichloran<br>Botran*<br>Ditranil | 2,6-Dichloro-4-nitroaniline | Selective fungicide used as soil treatment or foliage spray or dust; particularly effective against *Botrytis, Schlerotinia, Monilinia, Schlerotium* and *Rhizopus* | Tuco |
| **DCPA**<br>Dacthal* | Dimethyl-2,3,5,6-tetrachloroterephthalate | Pre-emergence herbicide for vegetables, cotton and ornamentals | Diamond |
| **DCU**<br>Dichloral urea<br>Crag* Herbicide 2 | 1,3-bis(1-Hydroxy-2,2,2-trichloroethyl) urea | Experimental for pre-emergence treatment in the control of annual grasses in certain broad-leaved crops | Carbide |

| Pesticides common name and synonyms | Chemical name and formula | Major uses | Manufacturers |
|---|---|---|---|
| **D-D***<br>Nemex*<br>Vidden D<br>Telone | Mixture of 1,3-dichloropropene, 3,3-dichloropropene, 1,2-dichloropropane, 2,3-dichloropropene and related $C_3$ chlorinated hydrocarbons<br>$C_3H_4Cl_2$ and $C_3H_6Cl_2$ | Soil fumigant for cysts, root-knot, root lesion and other plant parasites, nematodes, symphylids and wireworms | Shell<br>Dow |
| **DDD**<br>TDE<br>Dichloro-diphenyl-dichloroethane | 1,1-Dichloro-2,2-bis(*p*-chlorophenyl) ethane 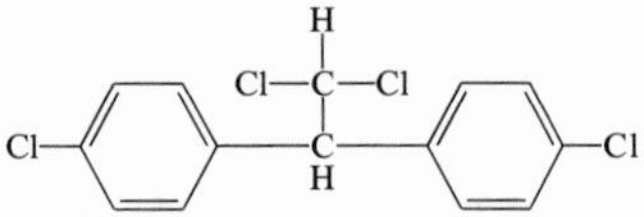 | Insecticide for fruits and vegetables against leaf-rollers, mosquito larvae, and hornworms | Rohm & Hass<br>Allied |
| **DDT**<br>Dichloro-diphenyl-trichloroethane<br>Genitox*<br>Anofex*<br>Chlorophenothane | 1,1,1-Trichloro-2,2-bis(*p*-chlorophenyl) ethane 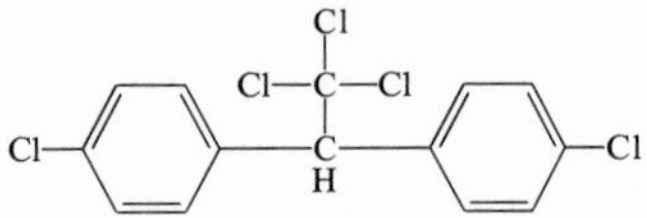 | Broad-spectrum insecticide for fruits, vegetables, cotton, household, livestock, timber, and industrial use | Lebanon<br>Allied<br>Diamond<br>Montrose<br>Olin<br>Geigy |
| **Def***<br>Degreen*<br>Ortho Phosphate Defoliant<br>Fos-Fall "A"<br>E-Z-Off D* | S,S,S-Tributyl phosphorotrithioate<br>$(CH_3-CH_2-CH_2-CH_2-S)_3-P{=}O$ | Defoliant in cotton | Chemagro<br>Chevron<br>Stauffer |

| | | | |
|---|---|---|---|
| **Dehydroacetic acid**<br>DHA | 2-Acetyl-hydroxy-3-oxo-4-hexenoic acid $\delta$-lactone | Mold preventive on processed foods; bananas, strawberries, squash | Dow |
| **Demeton**<br>Systox*<br>Mercaptophos | A 2:1 mixture of 0,0-diethyl-0-[2-(ethylthio)ethyl]phosphorothioate (thiono isomer) I and 0,0-diethyl-S-[2-(ethylthio)ethyl]phosphorothioate (thiolisomer) II<br>$(CH_3CH_2O)_2P(=S)-O-CH_2-CH_2-S-CH_2-CH_3$ I<br>$(CH_3CH_2O)_2P(=O)-S-CH_2-CH_2-S-CH_2-CH_3$ II | Systemic insecticide; foliage spray or soil soak in clover, barley, oats, wheat, fruit trees, hops, cotton, and sugar beets | Chemagro |
| **2,4-DEP**<br>Falone*<br>Falodin | tris[(2,4-Dichlorophenoxy)ethyl]phosphite<br>$[Cl-C_6H_3(Cl)-OCH_2CH_2O-]_3-P$ | Control of emerging weeds; does not kill established grasses or weeds; used on peanuts, corn, nonbearing strawberries, and potatoes at lay-by time | Uniroyal |
| **Dexon***<br>Bayer 22555 | *p*-Dimethylaminobenzenediazo sodium sulfonate<br>$(H_2C)_2N-C_6H_4-N=N-SO_2-ONa$ | Fungicide—protection of germinating seeds; used on sugar beets, corn, cotton, sorghum, avocados, beans, peas, spinach, cucumbers, ornamentals, and pineapples | Chemagro |

| Pesticides common name and synonyms | Chemical name and formula | Major uses | Manufacturers |
|---|---|---|---|
| **Diallate**<br>Avadex*<br>DATC<br>CP 15336 | S-2,3-Dichloroallyl diisopropylthio-carbamate 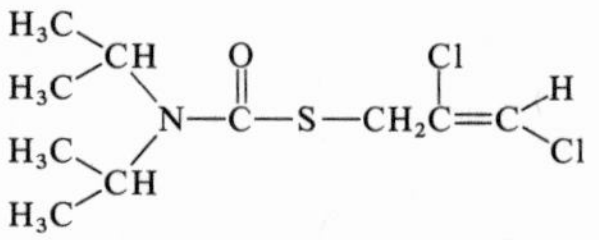 | Pre-emergence, selective herbicide for wild oats in sugar beets, corn, safflower | Monsanto |
| **Diazinon***<br>Alfa-tox*<br>Sarolex*<br>Basudin*<br>Spectracide*<br>AG 500 | 0,0-Diethyl 0-(2-isopropyl-4-methyl-6-pyrimidinyl) phosphorothioate 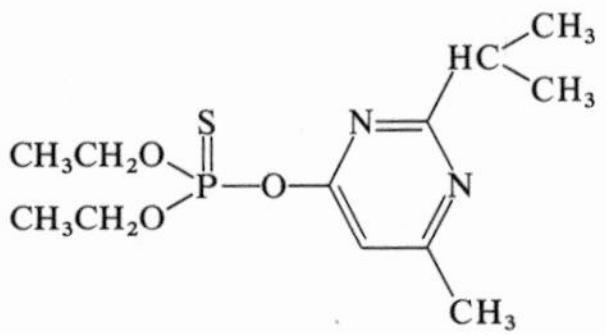 | For controlling resistant soil insects such as corn rootworm, wireworms, and cabbage maggot, also effective against many insect pests of fruits, vegetables, forage, field crops and ornamentals; also controls cockroaches and other household insects, nematodes, and insect pests of turf and lawns | Geigy |
| **Dicamba**<br>Banvel* D | 3,6-Dichloro-o-anisic acid 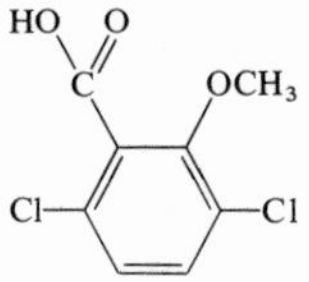 | For control of annual broad-leaved weeds in fall and spring, seeded small grains, established perennial grasses, golf course fairways and greens; pre- and post-emergence weed control in field corn, unwanted brush | Velsicol |

| | | | |
|---|---|---|---|
| **Dicapthon** | 0-2-Chloro-4-nitrophenyl 0,0-diethyl phosphorothioate 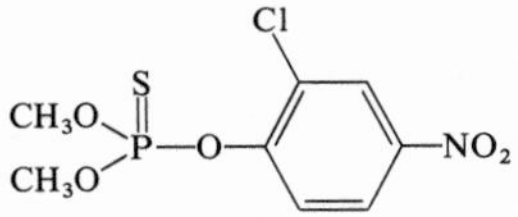 | Control of flies on farm buildings; effective against boll-weevil and roaches | Cyanamid |
| **Dichlobenil** Casoron* | 2,6-Dichlorobenzonitrile 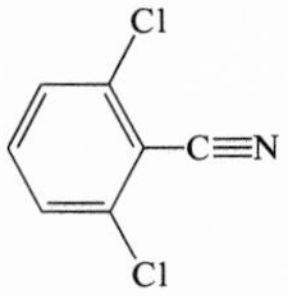 | Herbicide to treat cranberry bogs, fruit orchards, and ornamental nurseries | Thompson-Hayward |
| **Dichlone** Phygon* | 2,3-Dichloro-1, 4-naphthoquinone 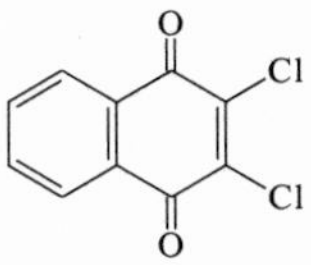 | Foliage treatment for fruits and vegetables; seed treatment for corn and beets; lake algicide | FMC Uniroyal |
| **Dichlorprop** 2-4 DP | 2-(2,4-Dichlorophenoxy)-propionic acid 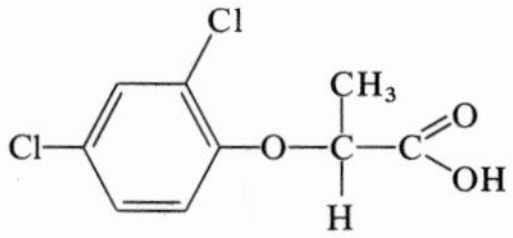 | Brush control on range lands and rights of way; aquatic weeds | Diamond Hercules |

| Pesticides common name and synonyms | Chemical name and formula | Major uses | Manufacturers |
|---|---|---|---|
| **Dichlorvos**<br>Vapona*<br>DDVP<br>Herkol*<br>Dedevap*<br>Oko*<br>Mafu* | 2,2-Dichlorovinyl dimethyl phosphate 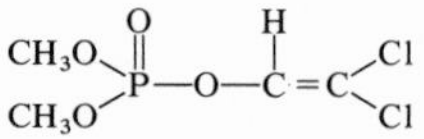 | Control of certain insects that are economically important in public health (man and livestock) and insects that attack stored products; effective against household pests | Shell |
| **Dicofol**<br>Kelthane | 1,1-bis(*p*-Chlorophenyl)-2,2,2-trichloroethanol 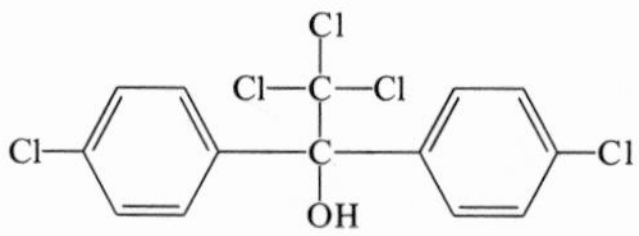 | No phytotoxic or insecticidal activity; used on various fruit and vegetable crops for mite control | Rohm & Haas |
| **Dieldrin**<br>Octalox*<br>Panoram D-31* | Not less than 85% of 1,2,3,4,10,10-hexachloro-6,7-epoxy-1,4,4a,5,6,7,8,8a-octahydro-1,4-*endo-exo*-5,8-dimethanonaphthalene 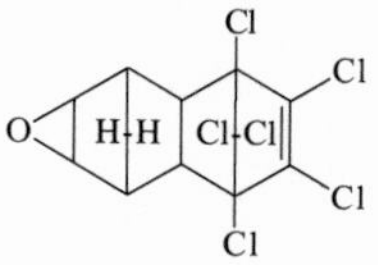 | To control general soil-inhabitating insects and certain insects attacking principal field, vegetable and fruit crops; mothproofing; public health pests; disease vectors; broad-spectrum insecticide | Shell |

| | | | |
|---|---|---|---|
| **Diethyltoluamide**<br>Deet<br>DET | N,N-diethyl-*m*-toluamide 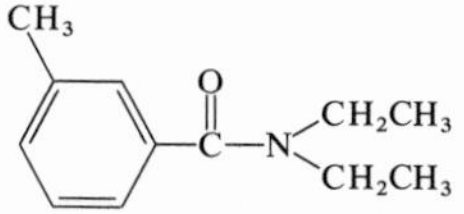 | Insect repellent for gnats, ticks, fleas, flies, mosquitoes, and chiggers | Chem. Form.<br>Hercules<br>Uniroyal |
| **Difolatan*** | *cis*-N-[(1,1,2,2-Tetrachloroethyl)thio]-4-cyclohexene-1,2-dicarboximide 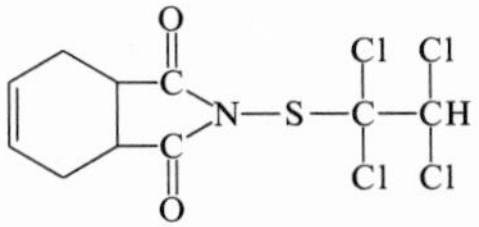 | Protectant fungicide for foliage application; controls early and late blight and late tuber rot on potatoes | Chevron |
| **Dilan***<br>Bulan*<br>Prolan* | Mixture containing 2-nitro-1,1-bis(*p*-chlorophenyl)butane (42.7%), 2-nitro-1,1-bis(*p*-chlorophenyl)propane (21.3%) and 16% related compounds in 20% xylene 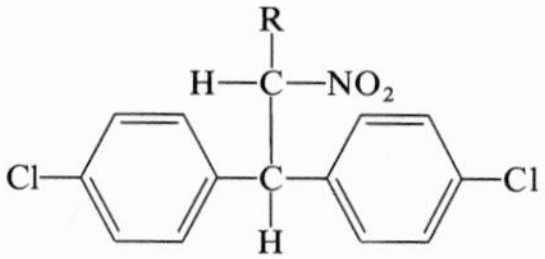<br>R=$CH_3CH_2$ (Bulan) or $CH_3$ (Prolan) | Insecticide for cotton, pears, beans, potatoes and for specific insect problems in the household; not approved for dairy or poultry use; no official tolerance | Comm. Solv. |
| **Dimefox**<br>Hanane<br>Pestox 14* | Tetramethyl phosphorodiamidic fluoride 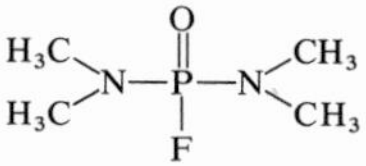 | Systemic acaricide; control of sap-feeding insects | Pest Control |

| Pesticides common name and synonyms | Chemical name and formula | Major uses | Manufacturers |
|---|---|---|---|
| **Dimethoate**<br>Cygon*<br>Rogor* | 0,0-Dimethyl S-(N-methylcarbamoyl-methyl) phosphorodithioate<br>$(CH_3O)_2P(=S)-S-CH_2-C(=O)-N(H)-CH_3$ | As a residual wall spray for controlling houseflies and for control of a wide range of insects on ornamental plants, certain vegetables, cotton, seed alfalfa, water-melons, bearing apples and pears, safflower, and nonbearing citrus | Cyanamid |
| **Dimethrin*** | 2,4-Dimethylbenzyl 2,2-dimethyl-3-(2-methylpropenyl)cyclopropane-carboxylate<br>$(H_3C)_2C=C(H)-C_3H_2(CH_3)_2-C(=O)-O-H_2C-C_6H_3(CH_3)_2$ | For human body lice, stable fly, hornfly, housefly, cattle lice, and face fly; used with synergist | MGK |
| **Dimethyl phthalate**<br>DMP | Dimethyl phthalate<br>$C_6H_4(C(=O)-O-CH_3)_2$ | Insect repellent for mosquitoes and chiggers | Allied |
| **Dimetilan**<br>Snip* Fly Bands | 2-Dimethylcarbamyl-3-methyl-5-pyrazolyl dimethylcarbamate<br>$(H_3C)_2N-C(=O)-N(\text{pyrazole},\ H_3C)-O-C(=O)-N(CH_3)_2$ | Impregnated in plastic fabric bands near ceiling in farm buildings for fly control | Geigy |

| | | | |
|---|---|---|---|
| **Dimite***<br>DCPC<br>DMC | 4,4′-Dichloro-α-methylbenzhydrol 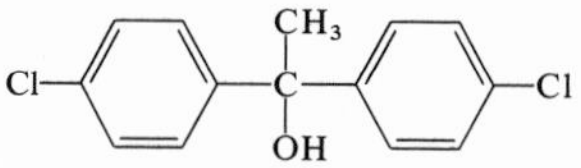 | Acaricide with ovicidal action; toxic to phytophagous mites, comparatively non-toxic to insects; used on beans, ornamentals, potatoes, and sugarbeets | Sherwin-Williams |
| **Dinitrocyclohexyl-phenol**<br>Dinex*<br>DN 111* (dicyclo-hexyl amine salt)<br>DN Dry Mix No. 1<br>DNOCHP | 4,6-Dinitro-o-cyclohexylphenol 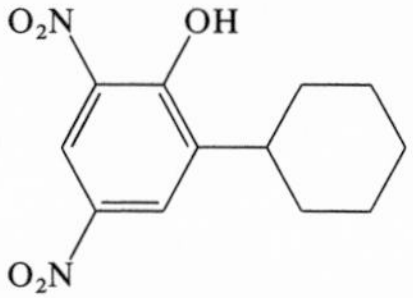 | Tree fruit protection against mites and insects | Chem. Ins. |
| **Dinobuton**<br>Dessin* | 2-sec-Butyl-4,6-dinitrophenyl isopropyl carbonate 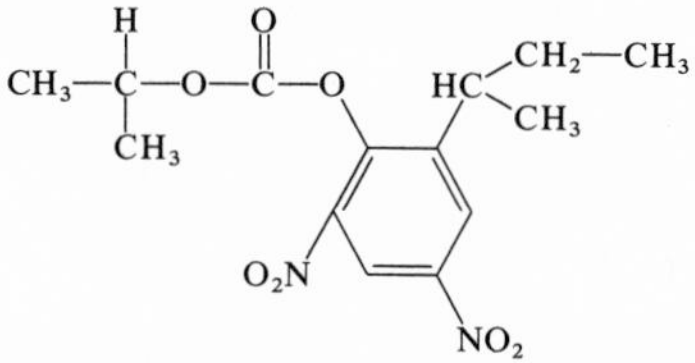 | Experimental on cotton, fruits, melons and other crops; effective against phytophagous mites and powdery mildew | Carbide |
| **Dinoseb**<br>DNBP<br>DNOSBP<br>Chem-Ox<br>Premerge* | 4,6-Dinitrophenol-o-sec-butylphenol 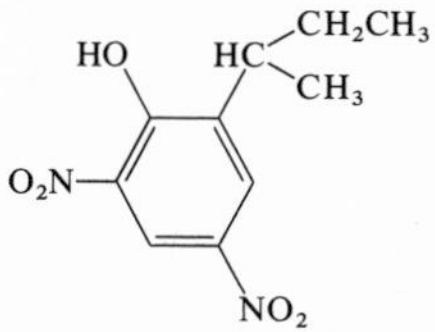 | Desiccant, flax, legumes, milo, and corn; weed control in vineyards and citrus groves | Chem. Ins.<br>Dow<br>FMC |

| Pesticides common name and synonyms | Chemical name and formula | Major uses | Manufacturers |
|---|---|---|---|
| **Dioxathion**<br>Delnav*<br>Navadel* | 2,3-*p*-Dioxanedithiol-S,S-bis (0,0-diethyl phosphorodithioate) 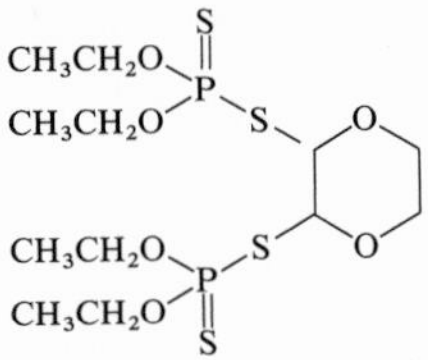 | Insecticide and acaricide relatively harmless to pollinating insects; control of mites on cotton grapes, citrus, ornamentals, apples, pears, quinces | Hercules |
| **Diphacinone**<br>Diphacin* | 2-Diphenylacetyl-1,3-indandione 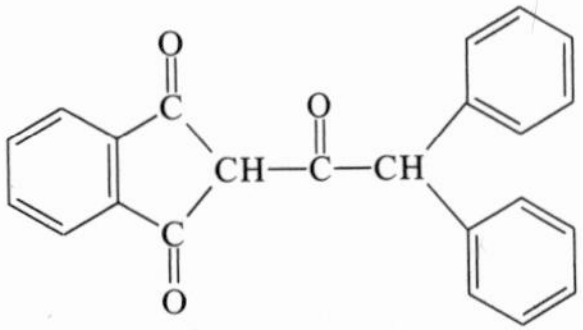 | Anticoagulant-type rodenticide | Nease |
| **Diphenamid**<br>Dymid*<br>Enide* | N,N-Dimethyl-2,2-diphenylacetamide 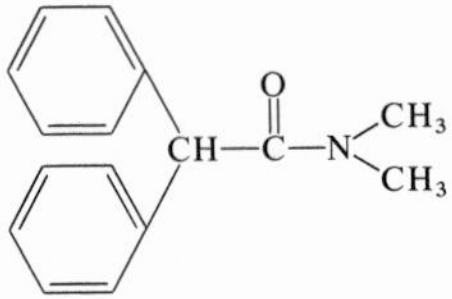 | Selective pre-emergence herbicide for control of annual grasses and certain broad-leaved weeds in direct seeded and transplant crops—e.g., tomatoes, potatoes, tobacco | Elanco<br>Tuco |

| | | | |
|---|---|---|---|
| **Diphenatrile**<br>Diphenylacetanitrile | Diphenylacetanitrile<br>$CH-C\equiv N$ | For pre-emergence control of seedling grasses in turf, flower beds, shrubbery planting, and broad-leaved ground covering | Elanco |
| **Diphenylamine** | Diphenylamine<br>NH | Screw-worm control in livestock; preharvest tree spray; postharvest spray or dip; impregnated wraps | Cyanamid<br>Dow<br>DuPont |
| **Diquat***<br>Reglone* | 6,7-Dihydrodipyrido[1,2a:2′,1′c] pyrazinedium salts<br>$\overset{+}{N}-\overset{+}{N}$ $2Br^-$ | Desiccation of seed crops; aquatic weed control | Chevron |
| **Disulfoton**<br>Disyston*<br>Dithiodemeton<br>Dithiosystox<br>Thiodemeton | 0,0-Diethyl-S-[2-(ethylthio)ethyl]-phosphorodithioate<br>$(CH_3CH_2O)_2P(=S)-S-CH_2-CH_2-S-CH_2-CH_3$ | Systemic insecticide for sucking and chewing insects on field, vegetable, fruit and nut crops | Chemagro |
| **Dithane M-45*** | Coordination product of zinc and manganese ethylene bisdithiocarbamate in which ingredients are zinc 2%, manganese 16% and ethylene bisdithio-carbamate 62% | Vegetable and fruit fungicide | Rohm & Haas |

| Pesticides common name and synonyms | Chemical name and formula | Major uses | Manufacturers |
|---|---|---|---|
| **Dithane S-31*** | Mixture of nickel sulfate and manganous ethylenebis[dithiocarbamate] | Foliage fungicide for control of rust on cereal grains and grasses | Rohm & Haas |
| **Diuron**<br>Karmex*<br>DMU<br>DCMU | 3-(3,4-Dichlorophenyl)-1,1-dimethylurea<br>$3,4\text{-}Cl_2C_6H_3-\underset{H}{N}-\overset{O}{\overset{\Vert}{C}}-N(CH_3)_2$ | For selective control of cotton, sugarcane, pineapple, grapes, alfalfa, apples, pears, citrus and other crops; general weed control on railroads, highways and industrial sites | DuPont |
| **DNOC**<br>Elgetol*<br>DNC<br>Sinox | 2-Methyl-4,6-dinitrophenol sodium salt<br>$NaO-C_6H_2(CH_3)(NO_2)_2$ (NaO, $CH_3$, $O_2N$, $NO_2$ on benzene ring) | Postemergence weed control in grains, flax, peas; dormant applications in fruits and walnuts; chemical thinning of apples, peaches, apricots, plums, prunes, and pears | FMC<br>Chem. Ins.<br>Dow |
| **Dodine**<br>Cyprex* | Dodecylguanidine acetate<br>$n-C_{12}H_{26}-\underset{H}{N}-C(=NH)NH_2 \cdot CH_3-\overset{O}{\overset{\Vert}{C}}OH$ | For control of scab on apples, pears, and pecans; leaf spot on cherries and peanuts; foliar diseases on strawberries; brown rot blossom blight on peaches and cherries; peach leaf curl; leaf blight of sycamores and black walnuts | Cyanamid |

| | | | |
|---|---|---|---|
| **DSMA**<br>Ansar* 184<br>Methar<br>DMA-100*<br>Crab-E-Rad 100*<br>Weed-E-Rad*<br>DiTac<br>MAA | Disodium methanearsonate<br>$\left[ CH_3-As(=O)(O^-)(O^-) \right] 2Na^-$ | Postemergence in cotton; general weed control in noncrop areas | Clearly<br>Diamond<br>Vineland<br>Ansul |
| **Dursban*** | 0,0-Diethyl-0-(3,5,6-trichloro-2-pyridyl)phosphorothioate<br>$(CH_3CH_2O)_2P(=S)-O-C_5NCl_3$ | Broad-spectrum insecticide, particularly mosquitoes, household pests and other soil insects | Dow |
| **Dyfonate***<br>N-2790<br>ENT-25,796 | 0-Ethyl-S-phenyl-ethylphosphonodithioate<br>$(CH_3CH_2O)(CH_3CH_2)P(=S)-S-C_6H_5$ | Insecticide for control of corn rootworm, wireworms, and symphylans | Stauffer |
| **Dyrene***<br>Kemate*<br>Triazine* | 2,4-Dichloro-6-(o-chloroaniline)-s-triazine<br>$C_3N_3Cl_2-NH-C_6H_4Cl$ | Turf fungicide; control of blight in potatoes and tomatoes, also other foliage applications—celery, onions | Chemagro |

| Pesticides common name and synonyms | Chemical name and formula | Major uses | Manufacturers |
|---|---|---|---|
| **EDB**<br>Ethylene dibromide<br>Bromofume*<br>Dowfume* W-85<br>Pestmaster* EDB-85<br>Soilbrom-85* | 1,2-Dibromoethane<br>$Br—CH_2—CH_2—Br$ | Effective stored-grain, space, and soil fumigant controlling both nematodes and insects | American Potash<br>Dow<br>FMC<br>Great Lakes<br>Michigan |
| **EGT**<br>Glytac* | Ethyleneglycol bis (trichloroacetate) 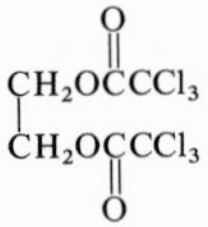<br>$CH_2OC(=O)CCl_3$<br>$\vert$<br>$CH_2OC(=O)CCl_3$ | Quack-grass control, also tall broadleaved weeds; cleared for use on cotton | Hooker |
| **Elcide* 73**<br>Merthiolate*<br>Thimerosal* | Sodium ethylmercurithiosalicylate 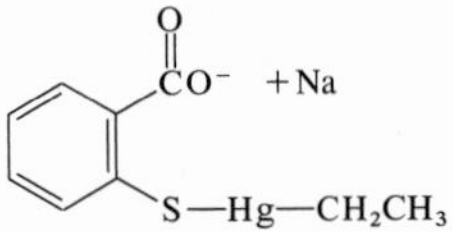<br>$C_6H_4(COO^-)(S—Hg—CH_2CH_3)$ +Na | Cotton seed treatment; gladiolus corms disease control | Elanco |
| **Emmi*** | 3,4,5,6,7,7-Hexachloro-N-(ethylmercuri)-1,2,3,6-tetrahydro-3,6-*endo*-methanophthalimide 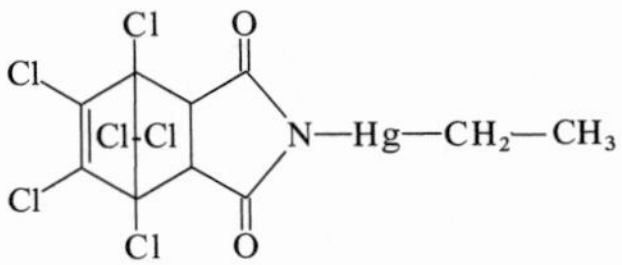<br>[structure: Cl, Cl, Cl, Cl, Cl–Cl ring with imide N—Hg—$CH_2$—$CH_3$] | Spore germination inhibitor for many fungi; used as seed dressing | Velsicol |

| | | | |
|---|---|---|---|
| **Endosulfan**<br>Thiodan* | 6,7,8,9,10,10-Hexachloro-1,5,5a,6,9,9a-hexahydro-6,9-methano-2,4,3-benzodioxathiepin-3-oxide 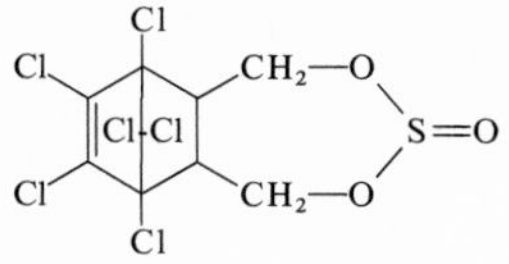 | Insect control on potatoes, cotton, seed peas, many other vegetables, tobacco, apples, peaches, strawberries, ornamentals | FMC |
| **Endothall**<br>Aquathol<br>Hydrothal<br>Accelerate<br>Des-I-Cate<br>Herbicide 282*<br>Herbicide 273* | 7-Oxabicyclo (2.2.1) heptane-2,3-dicarboxylic acid 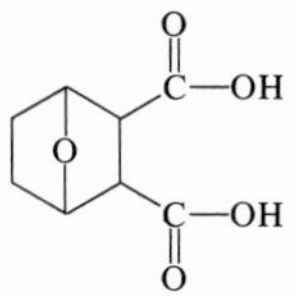 | Pre- and postemergence herbicide and harvest aid; root crop and vegetable protection; aquatic herbicide; alfalfa and clover desiccant | Pennsalt |
| **Endrin** | 1,2,3,4,10,10-Hexachloro-6,7-epoxy-1,4,4a,5,6,7,8,8a-octahydro-1,4-*endo*-*endo*-5,8-dimethanonaphthalene 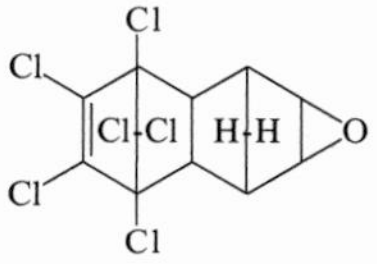 | Control of pests such as cotton insects, cutworms, armyworms, aphids, corn borer, cabbage looper, grasshoppers, plant bugs, lygus bugs, webworms and many other pests; also used as rodenticide | Shell<br>Velsicol |

| Pesticides common name and synonyms | Chemical name and formula | Major uses | Manufacturers |
|---|---|---|---|
| **EPN*** | O-Ethyl-O-*p*-nitrophenyl phenylphosphonothioate 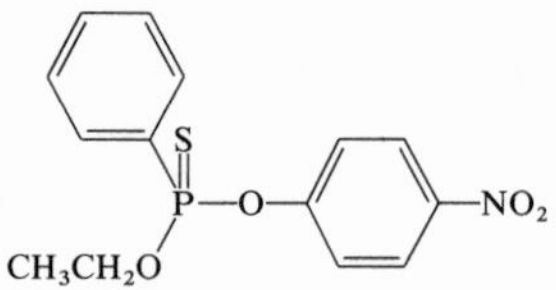 | To control mites and insects such as European corn borer, cotton boll weevil, plum curculio, Oriental fruit moth, rice stem borer | DuPont |
| **EPTC**<br>Eptam* | S-Ethyl dipropylthiocarbamate 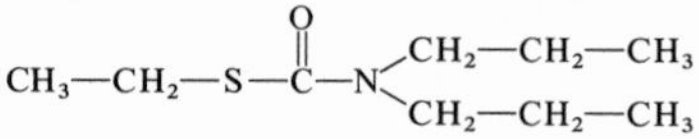<br>$CH_3-CH_2-S-C(=O)-N(CH_2-CH_2-CH_3)_2$ | Selective preemergence herbicide for use in dry and snap beans, Irish potatoes, alfalfa, corn, sugar beets, cotton and other; control of grassy weeds | Stauffer |
| **Eradex**<br>Thioquinox | 2,3-Quinoxalinedithiol cyclic trithiocarbamate 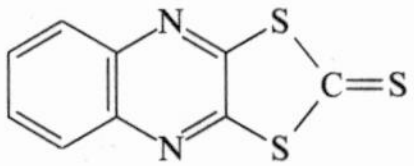 | Ovicide for spider mites; promising for control of powdery mildew | Bayer |
| **Erbon**<br>Baron*<br>Novege* | 2-(2,4,5-Trichlorophenoxy) ethyl 2,2-dichloropropionate 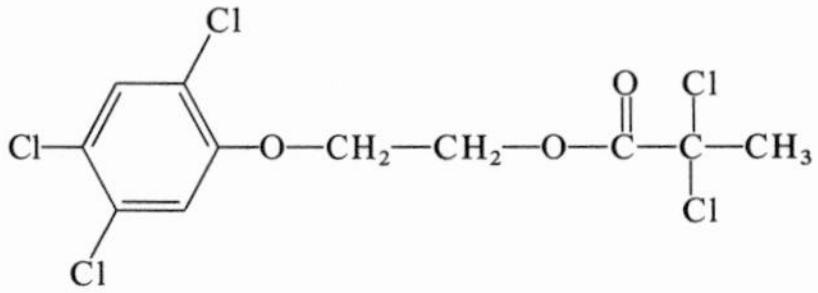<br>$Cl_3C_6H_2-O-CH_2-CH_2-O-C(=O)-CCl_2-CH_3$ | Systemic herbicide; nonselective and soil sterilant for one year; for noncrop areas only | Dow |

| | | | |
|---|---|---|---|
| **Ethide*** | 1,1-Dichloro-1-nitroethane 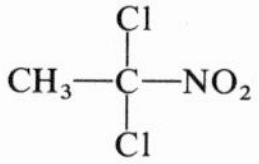 $CH_3-CCl_2-NO_2$ | Fumigation of stored grain, etc. | Comm. Solv. |
| **Ethion**<br>Nialate* | 0,0,0′,0′-Tetraethyl S, S′-methylene biophosphorodithioate 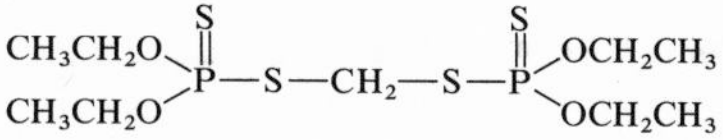 $(CH_3CH_2O)_2P(=S)-S-CH_2-S-P(=S)(OCH_2CH_3)_2$ | Controls overwintering stages of mites and aphids, control of phytophagous mites; used in combination with dormant oils against scales and as ovicide | FMC |
| **Ethoxyquin**<br>Stop-Scald* | 6-Ethoxy-1,2-dihydro-2,2,4-trimethylquinoline 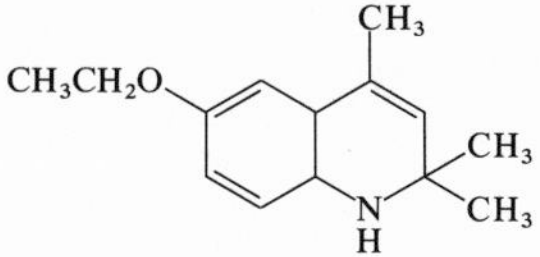 | Preharvest spray or postharvest dip for common scald of apples and pears in storage; fire retardant to prevent spontaneous combustion in high-fat-content meal; feed additive | Monsanto |
| **Ethylene** | Ethylene 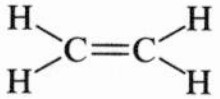 $H_2C=CH_2$ | Accelerates coloring of bananas, citrus, melons; blanching celery, endive and other vegetables | Carbide |
| **Ethylene dichloride**<br>EDC<br>Dowfume*<br>Chlorasol* | 1,2-Dichloroethane 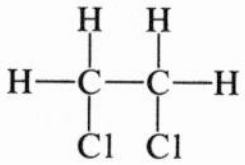 $H-CHCl-CHCl-H$ | Control of stored product pests, peachtree borer and Japanese beetle | Diamond<br>Dow<br>Carbide<br>Jefferson<br>PPG |
| **Ethylene oxide**<br>Oxirane | Ethylene oxide 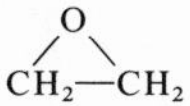 $CH_2-O-CH_2$ (ring) | Fumigant for stored foods | Dow<br>Jefferson<br>Carbide<br>Wyandotte |

| Pesticides common name and synonyms | Chemical name and formula | Major uses | Manufacturers |
|---|---|---|---|
| **Ethyl formate** | Ethyl formate<br>$H-C(=O)-OCH_2CH_3$ | Fumigant for food products; no deposit or residual odor; controls fruit beetle, saw-toothed grain beetle, raisin moth, grain beetle | Comm. Solv. |
| **Ethyl hexanediol**<br>6-12*<br>Ethohexadiol | 2-Ethyl-1,3-hexanediol<br>$CH_3CH_2CH_2CH(OH)CH(CH_2CH_3)CH_2OH$ | Repellent for most biting insects | Carbide |
| **Ethylmercury chloride**<br>Ceresan*<br>Granosan* | Ethylmercury chloride<br>$CH_3CH_2-Hg-Cl$ | For treatment of cotton, peanuts, and pea seeds to control numerous seed-borne diseases and to reduce seed decay and check damping-off; as a short-soak treatment for basal rot of narcissus bulbs | DuPont |
| **EXD**<br>Herbisan*<br>Bexide<br>Tridex | Diethyl dithiobis (thionoformate)<br>$S-C(=S)-O-CH_2-CH_3$<br>$S-C(=S)-O-CH_2-CH_3$ (S–S linked) | Pre-emergence herbicide; used for onions and also as preharvest desiccant | Roberts<br>Pennsalt |
| **Fenac***<br>Tri-Fene* | 2,3,6-Trichlorophenylacetic acid or sodium salt<br>$C_6H_2Cl_3-CH_2C(=O)-OH$ | Herbicide—pre-emergence for sugarcane, industrial areas, and aquatic weeds | Tenneco<br>Amchem |

| | | | |
|---|---|---|---|
| **Fenson***<br>Murvesco*<br>Trifenson* | *p*-Chlorophenylbenzenesulfonate 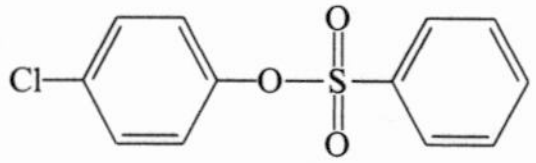 | Control of European red mite, brown mite (clover) by destroying eggs, immature spiders | Murphy |
| **Fenthion**<br>Baytex*<br>Bay 29493*<br>Entex*<br>Queletox*<br>Tiguvon*<br>Lebaycid | 0,0-Dimethyl-0-[4-(methylthio)-*m*-tolyl]-phosphorothioate 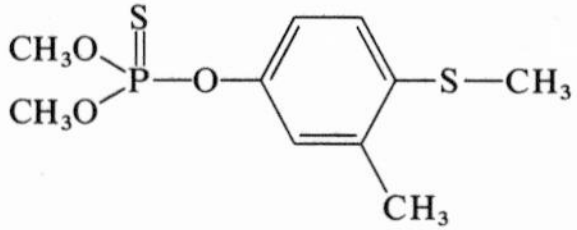 | Mosquitoes and larvae; flies, cockroaches, ants, fleas, crickets and wasps; Queletox* formulated for bird control | Chemagro |
| **Fenuron**<br>Dybar* | 3-Phenyl-1,1-dimethylurea 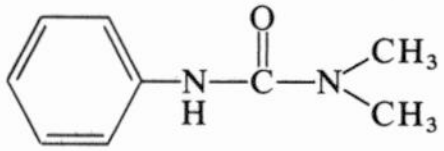 | Basal or broadcast soil treatment to control woody plant species; spot treatment for deep-rooted perennial weeds; general weed control on noncrop areas | DuPont |
| **Fenuron TCA**<br>Urab* | 3-Phenyl-1,1-dimethylurea trichloroacetate 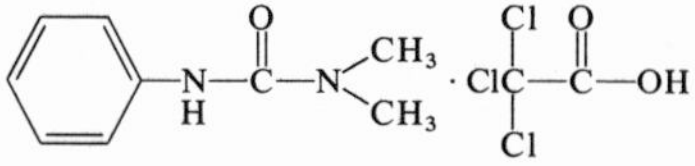 | Brush and perennial weed killer and temporary soil sterilant | Allied |
| **Ferbam**<br>Fermate*<br>Vancide-FE95 | Ferric dimethyl dithiocarbamate 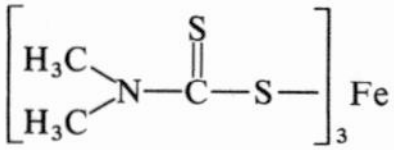 | To control many fungus diseases of fruits and nuts, certain vegetables, tobacco, and ornamentals | DuPont<br>FMC<br>Wood Ridge<br>Vanderbilt<br>Pennsalt |

| Pesticides common name and synonyms | Chemical name and formula | Major uses | Manufacturers |
|---|---|---|---|
| **Fluometuron**<br>Cotoran* | 1,1-Dimethyl-3-(α,α,α-trifluoro-*m*-tolyl) urea 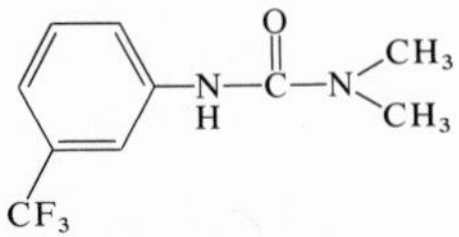<br>$CF_3C_6H_4-NH-C(=O)-N(CH_3)_2$ | Selective herbicide for weed control in cotton, asparagus, celery; sugarcane, woody ornamentals, citrus, small grains, and other crops | Ciba |
| **Folex***<br>Merphos*<br>Deleaf Defoliant*<br>Easy Off-D* | S,S,S-Tributyl phosphorotrithioite 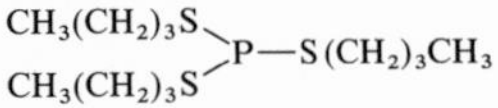<br>$[CH_3(CH_2)_3S]_2P-S(CH_2)_3CH_3$ | Cotton defoliant | Mobil |
| **Folpet**<br>Phaltan<br>Thiophal | N-(Trichloromethylthio)phthalimide 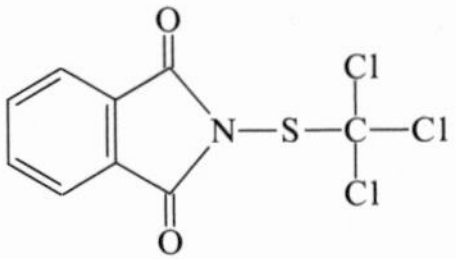<br>$C_6H_4(CO)_2N-S-CCl_3$ | Protectant eradicant fungicide for fruits, vegetables, flowers, ornamentals; control of apple scab, cherry leaf spot, rose black spot, rose mildew, and other plant diseases; seed and plant bed treatment | Stauffer<br>Chevron |
| **Formaldehyde**<br>Formalin | Formaldehyde 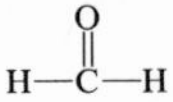<br>$H-C(=O)-H$ | Fumigant; soil sterilant (mushrooms) and seed treatment | Allied<br>Celanese |
| **Fumarin***<br>Coumafuryl | 3-(α-Acetonylfurfuryl)-4-hydroxycoumarin 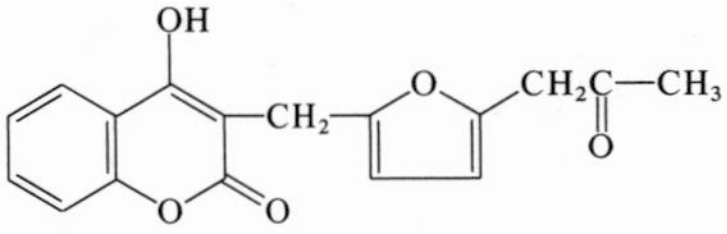<br>OH, $CH_2$, O, $CH_2C(=O)-CH_3$ | Anticoagulant rodenticide used as multiple-dose poison | Amchem |

| | | | |
|---|---|---|---|
| **Gardona***<br>Rabon* | 2-Chloro-1-(2,4,5-trichlorophenyl) vinyl dimethyl phosphate 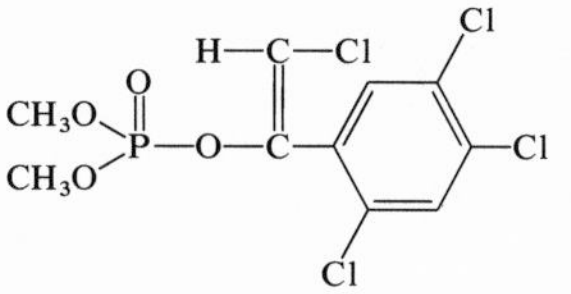 | Control of corn earworm and fall armyworm, codling moth, gypsy moth, houseflies and ectoparasites of livestock | Shell |
| **Genite (923)***<br>Genitol 923 | 2,4-Dichlorophenyl ester of benzenesulfonic acid 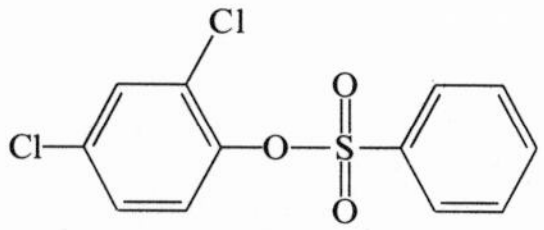 | Highly toxic to phytophagous mites; low insecticidal activity; used on fruit blossom | Allied |
| **Gibberellic acid**<br>Gibberellin | Gibberellic acid 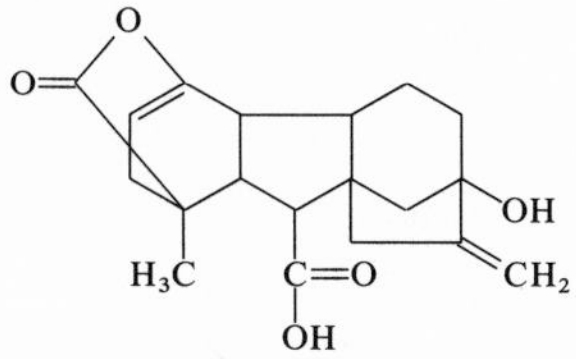 | Plant growth regulator, especially seed promotion | Amdal<br>Elanco<br>Merck<br>Pfizer |
| **Glyodin***<br>Crag* Fruit Fungicide<br>Glyoxide* | 2-Heptadecylimidazoline acetate 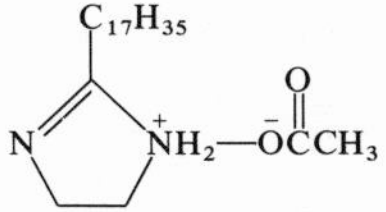 | Protective fungicide for scab, sooty blotch, Brooks spot, bitter rot, fly speck of apples, leaf spot of cherries, certain fungus diseases of ornamentals | PPG<br>Carbide |

| Pesticides common name and synonyms | Chemical name and formula | Major uses | Manufacturers |
|---|---|---|---|
| **Gophacide*** Bayer 38819 | 0,0-bis (*p*-Chlorophenyl) acetimidoylphosphoramidothioate 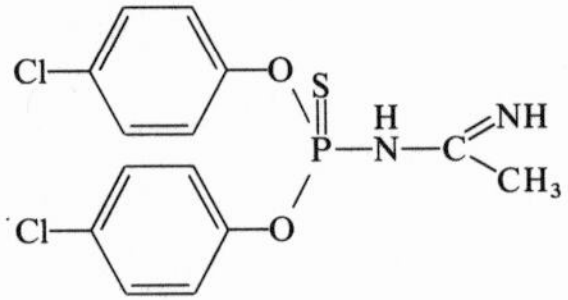 | Gopher bait | Chemagro |
| **Griseofulvin** | 7-Chloro-4,6-dimethoxy-coumaran-3-one-2-*spiro*-1′(2′-methoxy-6′-methylcyclohex-2′-en-4′-one) 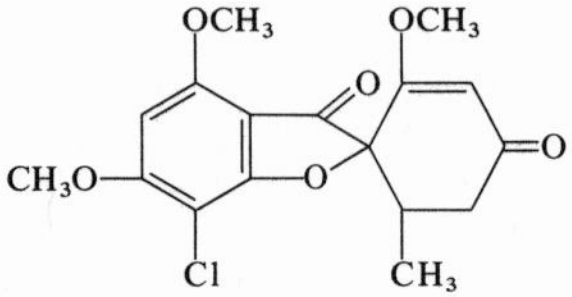 | Powerful fungicide—evidence of systemic activity | Merck Murphy |
| **HCA** Hexachloroacetone | Hexachloroacetone 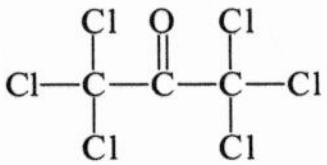 | Nonselective herbicide for pre- and post-emergence; Johnson grass, broad-leaved weeds | Allied |

| | | | |
|---|---|---|---|
| **HCB**<br>Hexachlorobenzene<br>Perchlorobenzene<br>No Bunt 40 or 80*<br>Anticarie* | Hexachlorobenzene 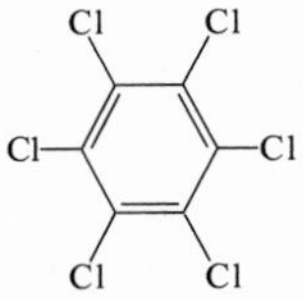 | Selective fungicide for seed treatment against wheat bunt | Chem. Ins. |
| **Heptachlor**<br>Drinox | 74% 1,4,5,6,7,8,8-heptachloro-3a,4,7a-tetrahydro-4,7-methanoindene 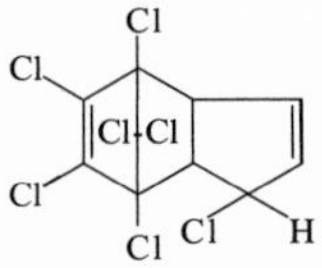 | Broad-spectrum insecticide against soil insects; for alfalfa, corn, cotton; control of soil insects in turf and field crops; termites, grasshoppers, household insects | Velsicol |
| **Hydrogen cyanide**<br>Hydrocyanic acid<br>Prussic acid | Hydrocyanic acid<br>H—C≡N | Fumigant for raw agricultural commodities, including grain | Cyanamid |
| **Hydroxymercuri-chlorophenol**<br>Semasan* | Hydroxymercurichlorophenol 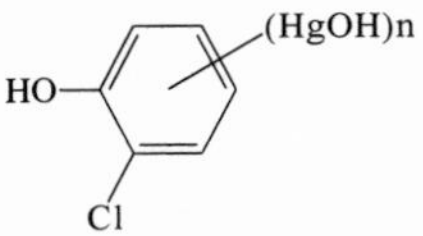<br>(HgOH)n, HO, Cl<br>n = 1 or 2 | Spray or dip treatment of Irish and sweet potatoes to control decay and surface-borne diseases; control of large brown patch, dollar spot, copper spot and other fine turf diseases | DuPont |
| **Igran 80W***<br>Prebane*<br>GS 14260 | 2-tert-Butylamino-4-ethylamino-6-methylthio-s-triazine 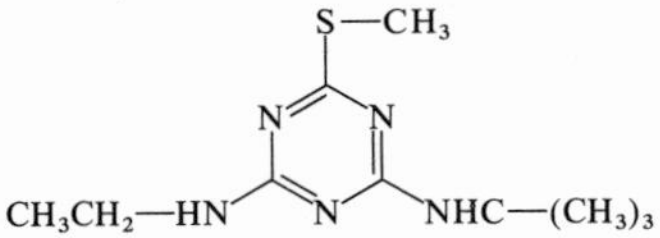<br>$S—CH_3$<br>$CH_3CH_2—HN$, $NHC—(CH_3)_3$ | Herbicide for control of most annual broad-leaved and grassy weeds in winter wheats, including many not controlled by 2, 4-D; experimental, full clearance expected later this year | Geigy |

| Pesticides common name and synonyms | Chemical name and formula | Major uses | Manufacturers |
|---|---|---|---|
| **Imidan*** <br> Prolate* | N-(Mercaptomethyl)phthalimide S-(0,0-dimethyl phosphorodithioate) 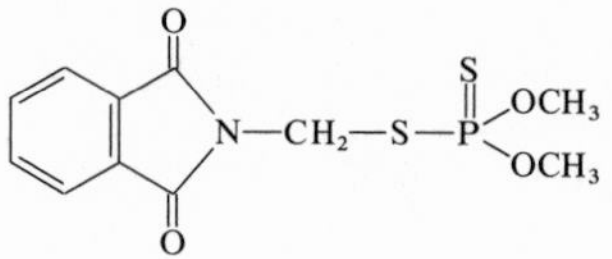 | Control of cotton boll weevil, codling moth | Stauffer |
| **Indolebutyric acid** <br> Hormodin* | Indole-3-butyric acid 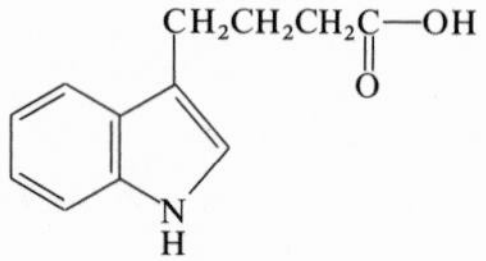 | Used in establishment of root cuttings | Merck |
| **Ioxynil** <br> Bantrol* <br> Actril* <br> Certrol* | 3,5-Diiodo-4-hydroxybenzonitrile 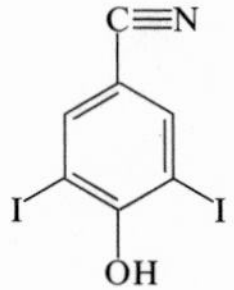 | Postemergence for seedling broadleaved weed control in cereal grains | Amchem <br> May & Baker <br> Chipman |

| Isocil<br>Hyvar*<br>Bromacil | 5-Bromo-3-sec-butyl-6-methyluracil 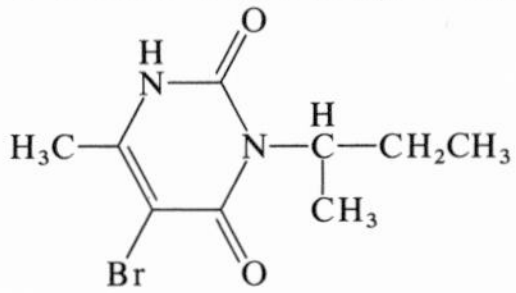 | Used as a general nonselective weed killer; particularly useful for control of perennial grasses | DuPont |
|---|---|---|---|
| Karathane*<br>Dinocap*<br>Arathane*<br>Mildex* | 2-(1-Methylheptyl)-4,6-dinitrophenyl crotonate 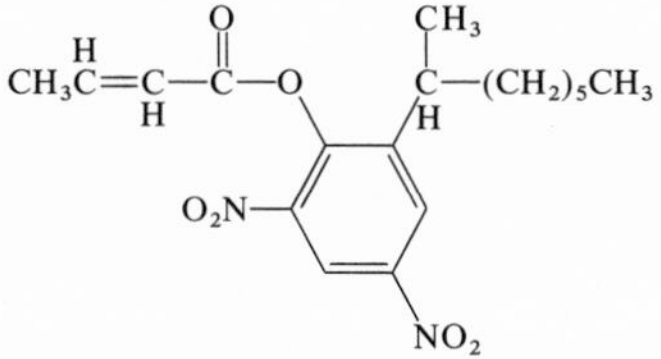 | Shows the acaricidal activity of the dinitrophenols; fruit and vegetable crop control of powdery mildew | Rohm & Haas |
| Kepone* | Decachlorooctahydro-1,3,4-metheno-2H-cyclobuta[cd] pentalen-2-one 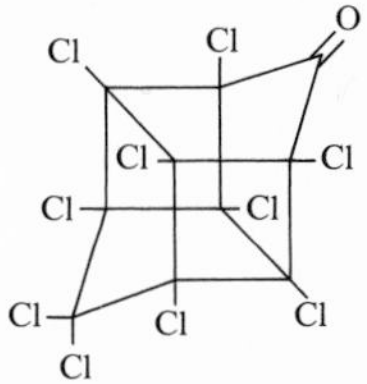 | Control of roaches, ants, wireworm, leaf-eating insects; fly larvacide | Allied |

| Pesticides common name and synonyms | Chemical name and formula | Major uses | Manufacturers |
|---|---|---|---|
| **KOCN**<br>Potassium cyanate<br>Aero* cyanate Weed Killer | Potassium cyanate<br>K—O—C≡N | Foliar spray in controlled emerged annual weeds in onions and certain other crops; control of crabgrass, chickweed, and annuals in turf | Cyanamid |
| **Landrin*** | 2,3,5-Trimethylphenyl methylcarbamate; 3,4,5-Trimethylphenyl methylcarbamate<br>$(CH_3)_3C_6H_2$—O—C(=O)—N(H)—$CH_3$ | Has performed well against corn root worm larvae, and is under study for a variety of soil and foliage insects | Shell |
| **Lanstan***<br>Chloronitropropane<br>NIA5961 | 1-Chloro-2-nitropropane<br>$CH_3$—CH($NO_2$)—$CH_2Cl$ | Soil fungicide for cotton, peas, beets, cucurbits | FMC |
| **Lasso*** | 2-Chloro-2′,6′-diethyl-N-(methoxymethyl)acetanilide<br>2,6-$(CH_2CH_3)_2C_6H_3$—N($CH_2OCH_3$)—C(=O)—$CH_2Cl$ | Control of certain annual grasses and broad-leaved weeds that infest soybean crops | Monsanto |

| | | | |
|---|---|---|---|
| **Lead arsenate** | Dibasic lead arsenate<br>$PbHAsO_4$ | Control of insects on apples, cherries, pears, peaches, grapes, celery, peppers, other fruits and vegetables | Allied<br>Chevron<br>Chipman<br>L.A. Chem.<br>FMC<br>Sherwin-Williams |
| **Lenacil**<br>Venzar* | 3-Cyclohexyl-6,7-dihydro-1H-cyclopentapyrimidine-2,4(3H,5H)-dione 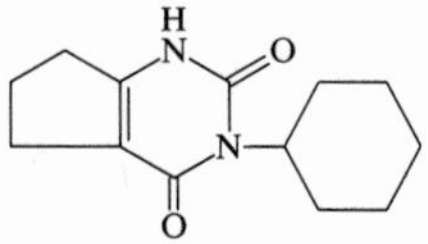 | In certain areas outside the U.S. for selective weed control in sugarbeets and strawberries | DuPont |
| **Lethane 384** | $\beta$-Butoxy-$\beta'$-thiocyanodiethyl ether<br>$CH_3(CH_2)_3{-}O{-}(CH_2)_2{-}O{-}(CH_2)_2{-}S{-}C{\equiv}N$ | Contact insecticide for flies on cattle and "knockdown" agent in homes; ovicidal to aphid eggs | Rohm & Haas |
| **Lime sulfur** | Aqueous solution of calcium polysulfides and calcium thiosulfate<br>not known | Dormant or delayed dormant application for diseases on apples, peaches, pears, and vegetables | Allied<br>Dow<br>Miller |
| **Lindane**<br>gamma BHC | 1,2,3,4,5,6-Hexachlorocyclohexane containing at least 99% gamma isomer 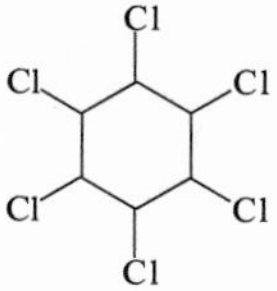 | Broad-spectrum insecticide for apples and other fruits, beans, peas, cole crops, cucurbits, tomatoes, other vegetables; also for dairy, livestock, household, and seed-treatment use | Diamond<br>Hooker |

| Pesticides common name and synonyms | Chemical name and formula | Major uses | Manufacturers |
|---|---|---|---|
| **Linuron**<br>Lorox*<br>Afalon | 3-(3,4-Dichlorophenyl)-1-methoxy-1-methylurea<br>$3,4\text{-}Cl_2C_6H_3\text{—NH—C(=O)—N(CH_3)—O—CH_3}$ | For selective weed control in corn, soybeans, grain sorghum, cotton, wheat, carrots, parsnips and potatoes; short-term control of annual weeds in noncropland areas such as roadsides and fencerows | DuPont |
| **Lovozal***<br>Fenzaflor<br>NC 5016 | Phenyl-5,6-dichloro-2-trifluoromethyl-benzimidazole-1-carboxylate<br>(5,6-Cl₂-benzimidazole ring, $CF_3$ at C-2, N-1—C(=O)—O—C₆H₅) | Experimental acaricide for control of summer mite infestations on apples | Fisons |
| **Magnesium chlorate**<br>E-Z Off*<br>Magron*<br>Ortho MC*<br>De-Fol-Ate* | Magnesium chlorate<br>$Mg(ClO_3)_2 \cdot 6H_2O$ | Defoliation of cotton, beans; sorghum desiccation | Allied<br>Chevron<br>Dow<br>Pennsalt |
| **Malathion**<br>Cythion*<br>Malaspray*<br>Chemathion* | 0,0-Dimethyl phosphorodithioate of diethyl mercaptosuccinate<br>$(CH_3O)_2P(=S)\text{—S—CH(C(=O)—O—CH_2CH_3)—CH_2—C(=O)—O—CH_2CH_3}$ | Controls many insects, including mosquitoes, houseflies, spider mites, aphids, scales, and a wide range of other sucking and chewing insects that attack fruits, vegetables, ornamentals, animals, and stored products | Cyanamid<br>Chem. Ins. |

| | | | |
|---|---|---|---|
| Maloran*<br>C-6313 | 3-(4-Bromo-3-chlorophenyl)-1-methoxy-1-methylurea 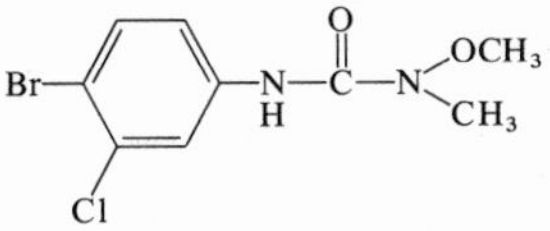 | For weed control in corn, safflower, sweet potatoes, carrots, celery, potatoes, small grains, and sugarcane | Ciba |
| MAMA<br>Ansar* 157 | Monoammonium methylarsenate 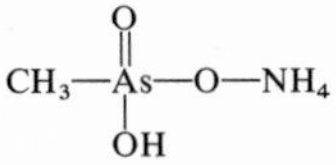 | Turf conditioner; removes crabgrass, nutgrass | Ansul |
| Maneb<br>Manzate*<br>MEB<br>MnEBD<br>Dithane M-22*<br>Chem Neb* | Manganous ethylene bisdithiocarbamate 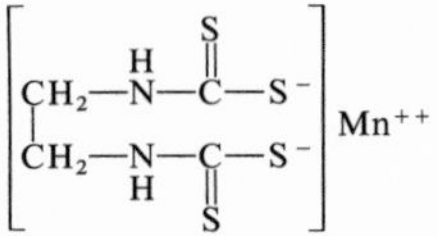 | To control many fungus diseases of vegetables, fruits and nuts, field crops, and ornamentals | DuPont<br>Chem. Ins.<br>Rohm & Haas<br>Pennsalt |
| Matacil*<br>Aminocarb<br>Bayer 44646 | 4-(Dimethylamino)-*m*-tolyl methyl carbamate 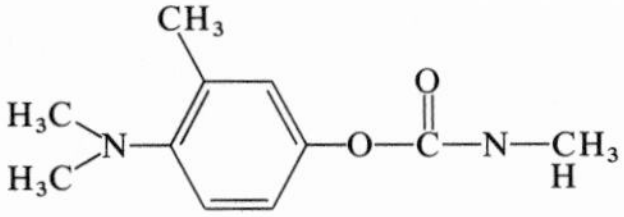 | Nonsystemic insecticide for lepidopterous larvae, other biting insects, boll weevil and bollworm on cotton; effective molluscide; some acaracidal activity | Chemagro |
| MCPA<br>MCP<br>Mephanac*<br>Methoxone*<br>Agroxone* | 2-Methyl-4-chlorophenoxyacetic acid 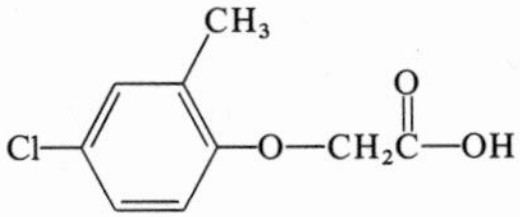 | Selective postemergence herbicide in grains, legumes, and corn | Monsanto<br>Dow<br>Chipman<br>Diamond |

| Pesticides common name and synonyms | Chemical name and formula | Major uses | Manufacturers |
|---|---|---|---|
| **MCPB**<br>Tropotox*<br>Thitrol*<br>Cantrol | 4-(2-Methyl-4-chlorophenoxy) butyric acid<br>$Cl-C_6H_3(CH_3)-O-(CH_2)_3-C(=O)-OH$ | More selective than MCPA; weed control in peas and clover | May & Baker<br>Dow<br>Monsanto<br>Amchem<br>Chipman |
| **MCPP**<br>Mecoprop<br>Mecopex*<br>Mecopar | 2-(2-Methyl-4-chlorophenoxy) propionic acid<br>$Cl-C_6H_3(CH_3)-O-CH(CH_3)-C(=O)-OH$ | Selective control of surface creeping broad-leaved weeds—e.g., clovers, chickweed, dandelion, ground ivy, knot weed, and plantain | Morton |
| **Mecarbam**<br>Murfotox*<br>Murotox*<br>Pestan | 0,0-Diethyl-S-(N-ethoxycarbonyl-N-methylcarbamoylmethyl)phosphorothiolothionate<br>$(CH_3CH_2O)_2P(=S)-S-CH_2-C(=O)-N(CH_3)-C(=O)-OCH_2-CH_3$ | Semisystemic control of scale insects, olive and fruit fly; root fly larvae in vegetables | Murphy |

| | | | |
|---|---|---|---|
| Memmi* | 3,4,5,6,7,7-Hexachloro-N-(methylmercuri)-1,2,3,6-tetrahydro-3,6-*endo*-methanophthalimide 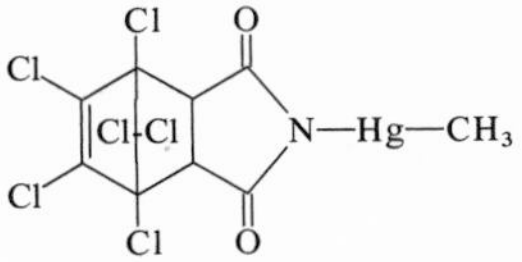 | Spore germination inhibitor; control of turf, corns, cereal grain and cotton seed disease | Velsicol |
| Menazon | S-(4,6-Diamino-s-triazin-2-ylmethyl) 0,0-dimethyl phosphorodithioate 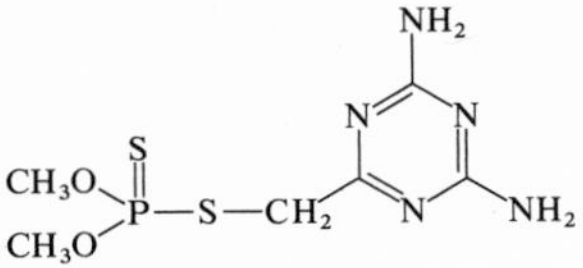 | Control of aphids, mites, leaf-hoppers; used as foliar spray, seed dressing or soil treatment | Plant Prot. |
| Mercuric chloride<br>Corrosive sublimate<br>Calo-Clor*<br>Calocure* | Mercuric chloride<br>$HgCl_2$ | Fungicide for seed, bulbs and tubers; seed potatoes; turf | Mallinckrodt |
| Mesurol*<br>Mercaptodimethur<br>Metmercapturon | 4-(Methylthio)-3,5-xylylmethylcarbamate 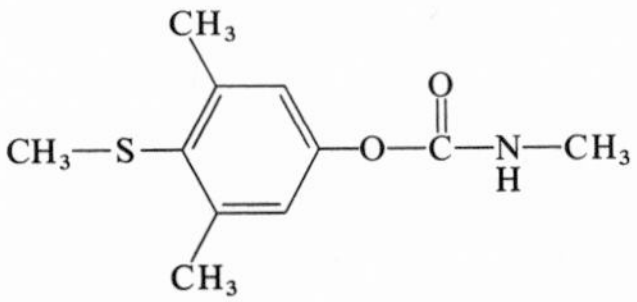 | Control of lepidopterous, coleopterous and dipterous pests; also has acaricidal activity | Bayer |
| Metaldehyde | Metaldehyde 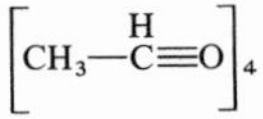<br>$\left[CH_3-\overset{H}{C}{\equiv}O\right]_4$ | Slug and snail control on vegetables | Comm. Solv. |

| Pesticides common name and synonyms | Chemical name and formula | Major uses | Manufacturers |
|---|---|---|---|
| **Metham**<br>SMDC<br>Vapam*<br>VPM<br>Metam | Sodium methyldithiocarbamate<br>$CH_3-N(H)-C(=S)-S-Na \cdot 2H_2O$ | Preplanting treatment to soil for control of certain weeds, soil-borne fungus diseases, nematodes, and garden centipedes | Stauffer<br>DuPont |
| **Methomyl**<br>Lannate* | S-Methyl N-[(methylcarbamoyl)oxy] thioacetimidate<br>$(CH_3S)(CH_3)C=N-O-C(=O)-N(H)-CH_3$ | Insecticide, nematocide | DuPont |
| **Methoxychlor**<br>Marlate*<br>DMDT<br>Methoxy DDT | 2,2-bis(*p*-Methoxyphenyl)-1,1,1-trichloroethane<br>$CH_3O-C_6H_4-CH(CCl_3)-C_6H_4-OCH_3$ | For control of a wide variety of insects attacking fruits, vegetables, field and forage crops, and livestock, as well as certain household and industrial insects | DuPont<br>Chem. Form. |
| **Methyl bromide**<br>Bromomethane<br>Dowfume* MC | Bromomethane<br>$CH_2Br$ | Insect control in elevators, mills, ships; greenhouse fumigation; soil fumigant for weeds, nematodes, insects, soil-borne diseases; termite control | American Potash<br>Dow<br>Great Lakes<br>Michigan<br>Vulcan |

| | | | |
|---|---|---|---|
| **Methyl-Demeton**<br>Meta-Systox*<br>Demeton-methyl<br>Demeton-S-methyl | 0-[2-(Ethylthio)ethyl]0,0-dimethyl phosphorothioate<br>$(CH_3O)_2P(=S)—O—CH_2—CH_2—S—CH_2—CH_3$<br>S-[2-(Ethylthio)ethyl]0,0-dimethyl phosphorothioate<br>$(CH_3O)_2P(=O)—S—CH_2—CH_2—S—CH_2—CH_3$ | Systemic insecticide for control of aphids and mites on corn, cotton, sugar beets, seed field crops, fruits, and vegetables | Chemagro |
| **Methyl parathion***<br>Metron*<br>Folidol M*<br>Meticide* | 0,0-Dimethyl 0-*p*-nitrophenyl phosphorothioate<br>$(CH_3O)_2P(=S)—O—C_6H_4—NO_2$ | For control of many insects of economic importance and is especially effective for boll-weevil control | Stauffer<br>Monsanto<br>Amer. Potash<br>Shell<br>Velsicol |
| **Methyl trithion** | S-[[(*p*-Chlorophenyl)thio]methyl] 0,0-dimethyl phosphorodithioate<br>$(CH_3O)_2P(=S)—S—CH_2—S—C_6H_4—Cl$ | Useful for control of a number of cotton insects, including boll-weevil | Stauffer |
| **Metobromuron**<br>Patoran* | 3-(*p*-Bromophenyl)-1-methoxy-1-methylurea<br>$(CH_3O)(CH_3)N—C(=O)—NH—C_6H_4—Br$ | For weed control in potatoes, tobacco, safflower, snap beans, and woody ornamentals | Ciba |

| Pesticides common name and synonyms | Chemical name and formula | Major uses | Manufacturers |
|---|---|---|---|
| **Mevinphos** Phosdrin* | 2-Carbomethoxy-1-methylvinyl dimethyl phosphate, α isomer 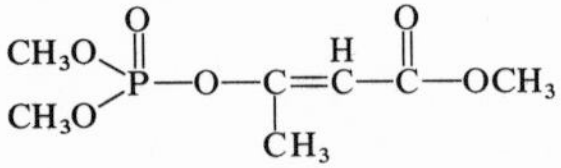 $(CH_3O)_2P(=O)-O-C(CH_3)=CH-C(=O)-OCH_3$ | Broad-spectrum, contact and systemic insecticide and acaricide giving rapid knockdown with short residual activity; controls certain economically important insects and mites that attack principal field, forage, vegetable, and fruit crops | Shell |
| **MGK 264*** Octacide 264 N-Octylbicycloheptenedicarboximide | N-(2-ethylhexyl)bicyclo[2.2.1]-5-heptene-2,3-dicarboximide 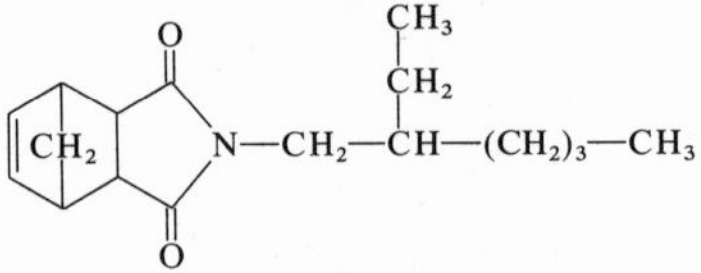 O, O, $CH_2$, $N-CH_2-CH(CH_2CH_3)-(CH_2)_3-CH_3$ | Synergist for pyrethrins, allethrin for household and industrial sprays | MGK |
| **MGK Repellent 11*** | 2,3:4,5-bis(2-Butylene)tetrahydro-2-furaldehyde 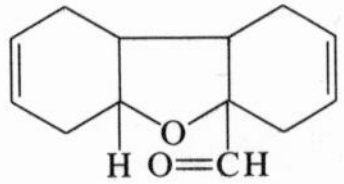 O, H, O=CH | Insect repellent stable against hornfly, roaches, mosquitoes; used on dairy cattle | MGK |
| **MGK Repellent 326*** | Dipropyl isocinchomeronate 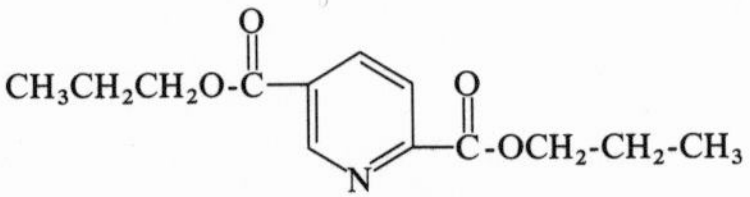 $CH_3CH_2CH_2O-C(=O)-C_5H_3N-C(=O)-OCH_2-CH_2-CH_3$ | Housefly repellent; dairy cattle spray | MGK |
| **MGK Repellent 874*** | 2-Hydroxyethyl-n-octyl sulfide $CH_3(CH_2)_7-S-CH_2CH_2OH$ | Residual roach and ant repellent | MGK |

| | | | |
|---|---|---|---|
| **MH***<br>Maleic hydrazide<br>MH-30*<br>Slo-Gro*<br>Sucker-Stuff*<br>Retard* | 1,2-Dihydro-3,6-pyridazinedione 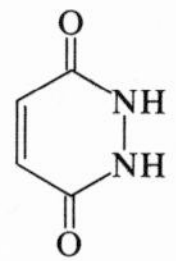 | Plant growth inhibitor; blocks cell division for inhibition of grasses, suckers on tobacco; sprout inhibitor for potatoes, onions | Uniroyal<br>Ansul<br>Chem. Form. |
| **Mirex**<br>Dechlorane | Dodecachlorooctahydro-1,3,4-metheno-1H-cyclobuta[cd]-pentalene 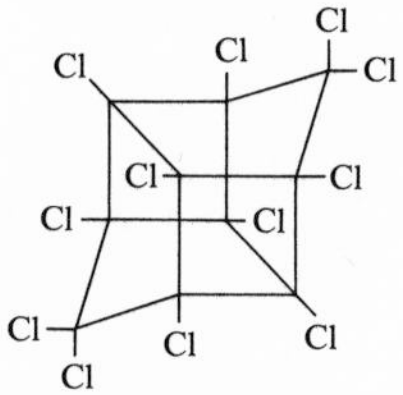 | Baits for control of fire ant, harvester ant and other species | Allied |
| **Mobam***<br>MCA 600* | 4-Benzothienyl-N-methylcarbamate 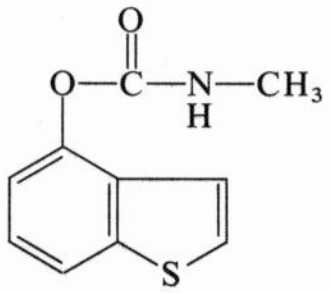 | Contact insecticide effective against cockroaches, flies, mosquitoes, aphids, grasshoppers, and a variety of crop insects | Mobil |
| **Molinate**<br>Ordram*<br>R-4572 | S-Ethyl hexahydro-1H-azepine-1-carbothioate 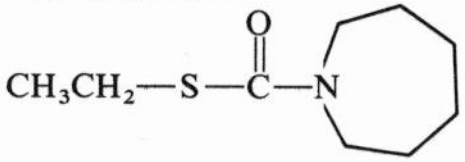 | Controls many germinating annual broad-leaved weeds and weed grasses, including wild oats | Stauffer |

| Pesticides common name and synonyms | Chemical name and formula | Major uses | Manufacturers |
|---|---|---|---|
| **Monitor*** Ortho 9006 ENT 27,396 | O,S-Dimethyl phosphoramidothioate $(CH_3O)(CH_3S)P(=O)NH_2$ | Broad-spectrum contact and residual acaracide; systemic activity for use on deciduous trees | Chevron |
| **Monuron** Telvar* | 3-(*p*-Chlorophenyl)-1,1-dimethylurea $Cl-C_6H_4-NH-C(=O)-N(CH_3)_2$ | For selective control of weed seedlings in asparagus, avocado, citrus fruits, cotton-seed, grapes, onions, pineapple, spinach | DuPont |
| **Monuron TCA** Urox* | 3-(4-Chlorophenyl)-1,1-dimethylurea trichloroacetate $Cl-C_6H_4-NH-C(=O)-N(CH_3)_2 \cdot Cl-CCl_2-C(=O)-OH$ | Nonselective soil sterilant, weed control | Allied |
| **Morestan*** Quinomethionate Oxythioquinox Bayer 36205 | 6-Methyl-2,3-quinoxalinedithiol cyclic-S, S-dithiocarbonate | Residual control of mites, mite eggs, powdery mildew, and pear psylla | Chemagro |
| **MPMT** Lambast* | 2,4-bis[(3-Methoxypropyl)-amino]-6-methylthio-s-triazine | Postemergence application to kill lamb's-quarters, pigweed, chickweed, etc., in safflower | Monsanto |

| | | | |
|---|---|---|---|
| **MSMA**<br>Diamond Arsonate Liquid<br>Ansar* 170, 529 | Monosodium acid methanearsonate<br>$CH_3-As(=O)(OH)-ONa$ | Control of Johnson grass and other grassy weeds in cotton, on ditch banks, rights-of-way, storage yards, and other noncrop areas; crabgrass control | Diamond<br>Ansul<br>Vineland |
| **Nabam**<br>Chem Bam*<br>Dithane D-14*<br>Dithane A-40* | Disodium ethylenebisdithiocarbamate<br>$CH_2-NH-C(=S)-S^-Na^+$<br>$\vert$<br>$CH_2-NH-C(=S)-S^-Na^+$ | Fungicide for foliage of potatoes, corn, celery, cole crops, beans, cucurbits, carrots, onions, tobacco, grapes, pecans, certain other fruits and vegetables; ornamentals | Chem. Ins.<br>FMC<br>Rohm & Haas |
| **Naled**<br>Dibrom* | 1,2-Dibromo-2,2-dichloroethyl dimethylphosphate<br>$(CH_3O)_2P(=O)-O-CH(Br)-CBr(Cl)-Cl$ | Contact insecticide and acaricide, with some fumigant action; brief residual activity | Chevron |
| **Naphthalene** | Naphthalene<br>$C_{10}H_8$ | Clothes moth, carpet beetle and household insects | Allied<br>Cyanamid<br>DuPont<br>Koppers<br>Reilly<br>Sherwin-Williams |
| **Naphthalene acetamide**<br>Rootone* | 1-Naphthalene acetamide<br>$C_{10}H_7-CH_2C(=O)-NH_2$ | Establishment of root cuttings, thinning of apples and pears | Amchem<br>Thompson-Hayward |

| Pesticides common name and synonyms | Chemical name and formula | Major uses | Manufacturers |
|---|---|---|---|
| **Naphthaleneacetic acid** | 1-Naphthaleneacetic acid 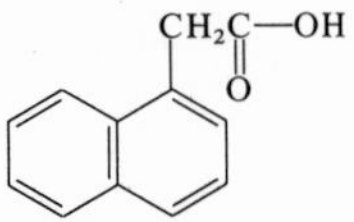 | Useful in rooting plants, including pineapple flowering, and controlling fruit set | Amchem<br>Chem. Form<br>Thompson-Hayward |
| **Naptalam**<br>Alanap*<br>NPA<br>Dyanap (mixed with DNBP) | N-1-Naphthylphthalamic acid 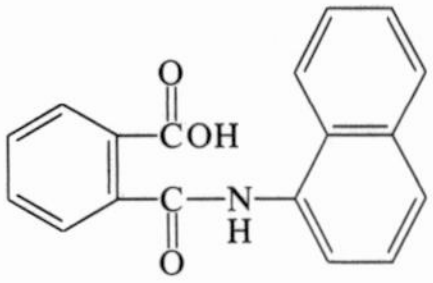 | Selective pre-emergence herbicide for soybeans, peanuts, sweet potatoes, Irish potatoes, cucurbits | Uniroyal |
| **Neburon**<br>Kloben* | 1-Butyl-3-(3,4-dichlorophenyl)-1-methylurea 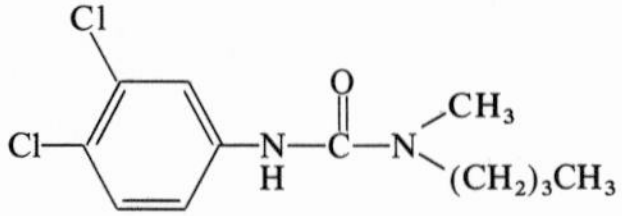 | Controls many germinating broad-leaved weeds and weed grasses | DuPont |
| **Nemacide**<br>V-C-13* | 0-2,4-Dichlorophenyl 0,0-diethyl phosphorothioate 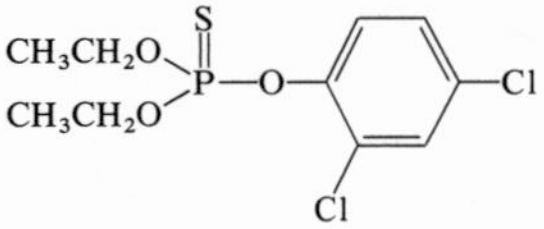 | Nematocide on turf, ornamentals, corn, cucumber, pepper, strawberry, squash, ornamentals, potting soil | Mobil |

| | | | |
|---|---|---|---|
| **Neo-Pynamin***<br>Phthalthrin | 2,2-Dimethyl-3-(2-methylpropenyl) cyclopropanecarboxylic ester of N-(hydroxymethyl)-1-cyclohexene-1,2-dicarboximide 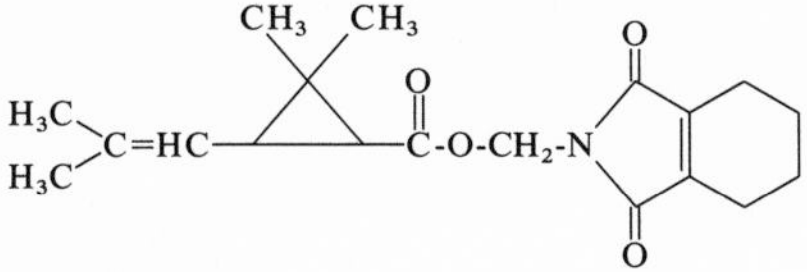 | Household insecticide for the control of flying and crawling insects | Sumitomo |
| **NIA 10637** | Ethyl hydrogen 1-propyl phosphonate 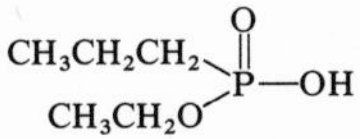 | Experimental plant growth regulator for control of woody perennials and annuals; tree and brush control | FMC |
| **Niacide*** | Mixture of manganese dimethyl dithiocarbamate and mercaptobenzothiazole 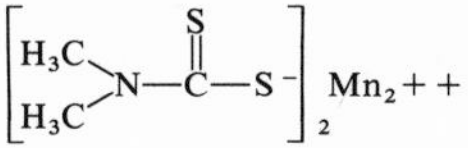 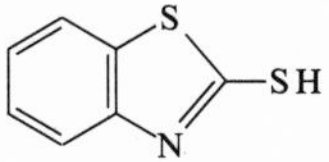 | Fungicide for apple foliage | FMC |
| **Nicotine**<br>Black Leaf Products<br>Black Leaf 40*<br>Nicotine sulfate | 3-(1-Methyl-2-pyrrolidyl)pyridine 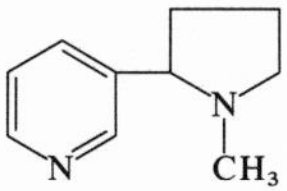 | Insecticide on apples, apricots, strawberries, cucurbits, certain other fruits and vegetable crops, ornamentals and home gardens | Chem. Form. |

| Pesticides common name and synonyms | Chemical name and formula | Major uses | Manufacturers |
|---|---|---|---|
| **Nitralin**<br>Planavin* | 4-(Methylsulfonyl)-2,6-dinitro-N,N-dipropylaniline 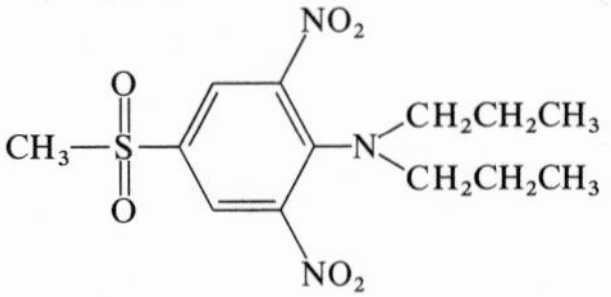 | Selective pre-emergence herbicide for weed control in cotton and soybeans; controls annual grasses and many broad-leaved weeds | Shell |
| **Nitrofen**<br>TOK E-25<br>NIP<br>F-W-925 | 2,4-Dichlorophenyl *p*-Nitrophenyl ether 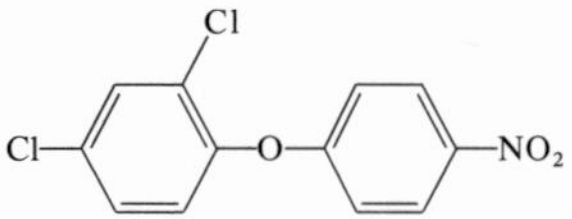 | Selective postemergence herbicide for control of broadleaved weeds and grasses in rice, cole, nutgrass, and other crops | Rohm & Haas |
| **Norbormide**<br>Raticate*<br>Shoxin* | 5-(α-Hydroxy-α-2-pyridylbenzyl)-7-(α-2-pyridylbenzylidene)-5-norbornene-2,3-dicarboximide 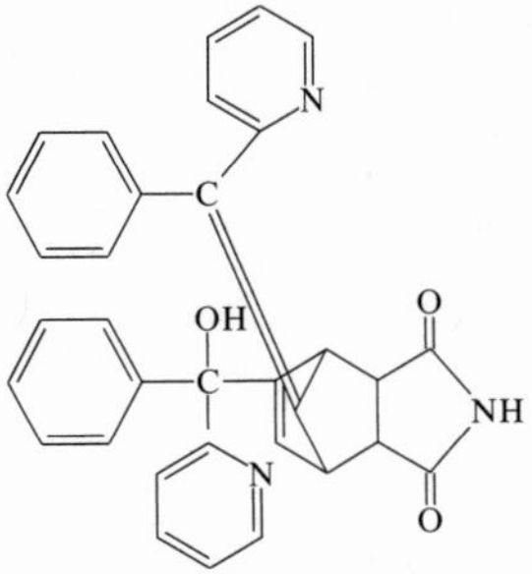 | Selective rodenticide; nontoxic to most other mammals; acts on rat's circulatory system | J & J |

| | | | |
|---|---|---|---|
| **Norea**<br>Herban*<br>Hercules 7531 | 3-(Hexahydro-4,7-methanoinden-5-yl)-1,1-dimethylurea 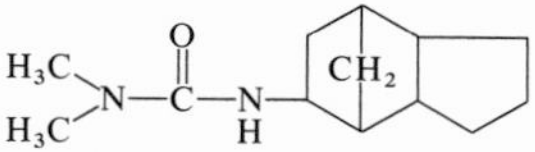 | Control of annual grasses; pre-emergence treatment of cotton, sugarcane, etc. | Hercules |
| **Orthodichloro benzene** | o-Dichlorobenzene 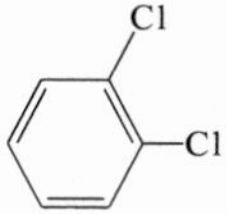 | Fumigant against household pests in low-pressure aerosols; termite control | Dow<br>Monsanto<br>Hooker |
| **Ortho LM**<br>LM seed protectant | Methyl mercury 8-hydroxyquinolate 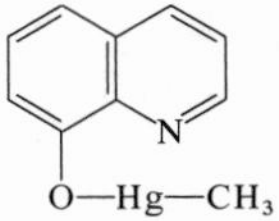 | Seed treatment for wheat, barley, oats, rice, flax, cotton, and soybeans | Chevron |
| **Orthophenylphenol**<br>OPP<br>Ortho-xenol<br>Dowicide A*<br>Topane*<br>RCI 49-155<br>Sodium-o-phenylphenate<br>2-Hydroxybiphenyl | 2-Phenylphenol 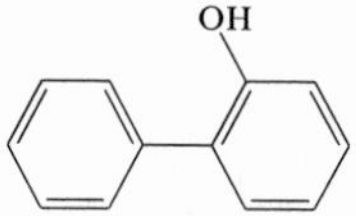 | Postharvest application in wax applied commercially to fruit and vegetables; dip for crates, hampers, etc.; industrial preservative | Reichold<br>Dow |
| **Ovex**<br>Ovotran*<br>Chlorfenson<br>Chlorofenson | *p*-Chlorophenyl-*p*-chlorobenzenesulfonate 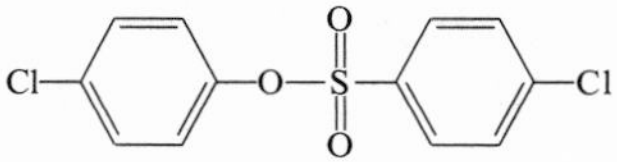 | Mite ovicide, also toxic to active stages; tree fruits, nuts, cotton, melons, strawberries | Dow |

| Pesticides common name and synonyms | Chemical name and formula | Major uses | Manufacturers |
|---|---|---|---|
| **Panogen***<br>Pandrinox*<br>Morsodren*<br>Panodrin A 13*<br>Panterra* | Methyl mercury dicyandiamide<br>$CH_3{-}Hg{-}\overset{H}{N}{-}C({=}NH){-}\underset{H}{N}{-}C{\equiv}N$ | Seed dressing for small grain, cotton, rice, flax, safflower, sorghum, beans, peas, soybeans; dip for corms, bulbs, potatoes; turf | Morton |
| **Paradichlorobenzene** | *p*-Dichlorobenzene<br>$Cl{-}C_6H_4{-}Cl$ | Fumigant against household insects in mothballs; peach tree borers, bark beetles; tobacco blue mold, mildew and other fungi | Allied<br>Monsanto<br>Dow<br>PPG |
| **Paraquat**<br>*Ortho*-paraquat chloride<br>*Ortho*-dual paraquat | 1,1′-Dimethyl-4,4′-bipyridinium dichloride or 1,1′-Dimethyl-4,4′-bipyridinium-bis [methylsulfate]<br>$CH_3{-}\overset{+}{N}C_5H_4{-}C_5H_4\overset{+}{N}{-}CH_3$<br>$2Cl^-$ or $2CH_3SO_4^-$ | Herbicide for weeds and grasses for crop and noncrop areas. Harvest aid chemical for cotton and potatoes | Chevron |
| **Parathion***<br>Phoskil*<br>Thiophos*<br>Folidol<br>E-605<br>Niran<br>Alleron*<br>Alkron<br>Ethyl parathion<br>Corothion*<br>Orthophos<br>Panthion*<br>Parawet*<br>Stathion* | 0,0-Diethyl 0-*p*-nitrophenyl phosphorothioate<br>$(CH_3CH_2O)_2P({=}S){-}O{-}C_6H_4{-}NO_2$ | Broad-spectrum insecticide effective against aphids, mites, *Lepidoptera*, beetles, leaf-hoppers and thrips on fruits, vegetables, and forage crops; cotton insects, symphilids, rootworms and other soil insects | Stauffer<br>Cyanamid<br>Monsanto<br>Shell<br>Velsicol<br>Amer. Potash |

| | | | |
|---|---|---|---|
| **Paris green**<br>Copper acetoarsenite | Copper *meta*-arsenite copper acetate complex<br>$(CH_3-\overset{O}{\overset{\parallel}{C}}-O)_2-Cu \cdot 3Cu(AsO_2)_2$ | Foliage applications in potatoes, sugar beets, cabbage; bait application in field, vegetable, and fruit crops; mosquito larvicide | Chipman<br>L.A. Chem.<br>Sherwin-Williams |
| **PAS**<br>Puratized*<br>agricultural spray | Phenylmercuritriethanolammonium lactate<br>$C_6H_5-Hg-\overset{+}{N}-(C_2H_4OH)_3$<br>$^{-}O-\underset{O}{\underset{\parallel}{C}}-\underset{OH}{\underset{\mid}{CH}}-CH_3$ | Fungicide for ornamentals and turf only | FMC |
| **PBA**<br>Zobar*<br>Polychlorobenzoic acid | Dimethylamine salt of polychlorobenzoic acid<br>$C_6H_{(5-n)}Cl_n-\overset{O}{\overset{\parallel}{C}}-OH \cdot HN(CH_3)_2$<br>Cl 3 or 4 | Bind weed and some deep-rooted perennial broad-leaved weeds; some species of woody plants | DuPont |
| **PCNB**<br>Pentachloronitro-benzene<br>Terraclor* | Pentachloronitrobenzene<br>$C_6Cl_5-NO_2$ | Soil fungicide for cotton, crucifers, potatoes, lettuce, peanuts, wheat, beans, tomatoes, peppers, and ornamentals | Olin |
| **PCP**<br>Pentachlorophenol<br>RCl 49–162<br>Sodium pentachlorophenate | Pentachlorophenol<br>$C_6Cl_5-OH$ | Contact herbicide and wood preservative; herbicide and desiccant on sugarcane; molluscicide to control snail carriers of larval human blood flukes causing schistosomiasis | Dow<br>Monsanto |

| Pesticides common name and synonyms | Chemical name and formula | Major uses | Manufacturers |
|---|---|---|---|
| **Pebulate**<br>Tillam*<br>PEBC | S-Propyl butylethylthiocarbamate<br>$CH_3CH_2CH_2{-}S{-}C({=}O){-}N{<}^{(CH_2)_3CH_3}_{CH_2-CH_3}$ | Controls both grassy and broad-leaved weeds; selective weed control in sugar beets and tobacco | Stauffer |
| **Penar*** | Dimethyldodecylamine acetate<br>$CH_3(CH_2)_{11}{-}\overset{+}{N}(CH_3)_2{-}H\cdot CH_3C({=}O){-}O^-$ | Plant growth regulator, to control tobacco suckers | Pennsalt |
| **Pentac*** | Decachlorobi(2,4-cyclopentadiene-1-yl)<br>$C_{10}Cl_{10}$ | Miticide for greenhouse floral crops and noncrop nursery stocks | Hooker |
| **Perthane***<br>Diethyl-diphenyl-dichloroethane | 1,1-Dichloro-2,2-bis(*p*-ethylphenyl) ethane (95%) plus related compounds<br>$(H_5C_2{-}C_6H_4)_2CH{-}CHCl_2$ | Vegetables and fruits, crucifers, lettuce, peas: household insects, dairy sprays | Rohm & Haas |
| **Petroleum oils**<br>Orchex* | Paraffinic base | Spray oils for deciduous and citrus fruit trees, nut trees, and ornamentals; non-phytotoxic oil as postemergence carrier for herbicides | Humble |

| | | | |
|---|---|---|---|
| **Phenothiazine**<br>Thiodiphenylamine<br>Phenthiazine | Dibenzo 1,4-thiazine | Dust for apple, pear, quince trees; control of hornfly and facefly; anthelmintic "wormer" fed in salt of mineral supplement | Atomic |
| **Phenylmercury urea**<br>Agrox* | Phenylmercury urea<br>$C_6H_5-Hg-NH-C(=O)-NH_2$ | Seed treatment on cotton, corn, rice and wheat | Chipman |
| **Phorate**<br>Thimet* | 0,0-Diethyl S-(ethylthio)-methyl phosphorodithioate<br>$(CH_3CH_2O)_2P(=S)-S-CH_2-S-CH_2-CH_3$ | Systemic insecticide on sugar beets, potatoes, lettuce, peanuts, corn, rice, barley, cotton, wheat, tomatoes, beans, ornamentals, cotton seeds | Cyanamid |
| **Phosfon*** | Tributyl 2,4-dichlorobenzylphosphonium chloride<br>$[Cl-C_6H_3(Cl)-CH_2-P-(C_4H_9)_3]^+ Cl^-$ | Height retardant for ornamentals | Mobil |
| **Phosphamidon**<br>Dimecron | 2-Chloro-2-diethylcarbamoyl-1-methylvinyl dimethyl phosphate<br>$(CH_3O)_2P(=O)-O-C(CH_3)=C(Cl)-C(=O)-N(CH_2CH_3)_2$ | Effective against aphids, mites, beetles and plant insects, both systemic and contact | Chevron |
| **Picloram**<br>Tordon* | 4-Amino-3,5,6-trichlóropicolinic acid | Controls several brush species; use on noncrop land, including utility rights-of-way and industrial storage areas | Dow |

| Pesticides common name and synonyms | Chemical name and formula | Major uses | Manufacturers |
|---|---|---|---|
| **Piperonyl butoxide**<br>Butacide* | Butylcarbityl-6-propylpiperonyl ether (80%) and related compounds (20%) 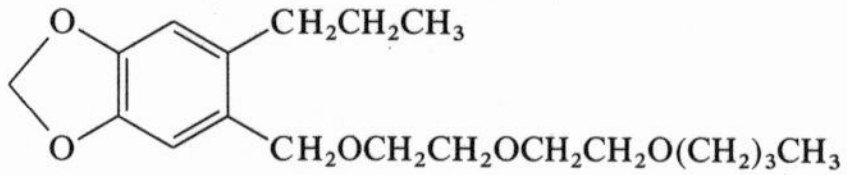<br>$CH_2CH_2CH_3$<br>$CH_2OCH_2CH_2OCH_2CH_2O(CH_2)_3CH_3$ | Synergist for pyrethrins and allethrin used to control household flying insects, stored grain pests, fruit fly on vegetables after harvest | FMC |
| **Pipron***<br>Piperalin | 3-(2-Methylpiperidino)propyl 3,4-dichlorobenzoate 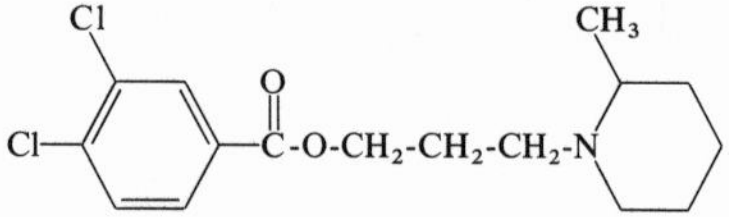<br>$CH_3$<br>Cl—C(=O)-O-$CH_2$-$CH_2$-$CH_2$-N | Controls or prevents powdery mildew in ornamentals | Elanco |
| **Pival***<br>Pindone<br>Pivalyn*<br>Tri-ban* | 2-Pivalyl-1,3-indandione 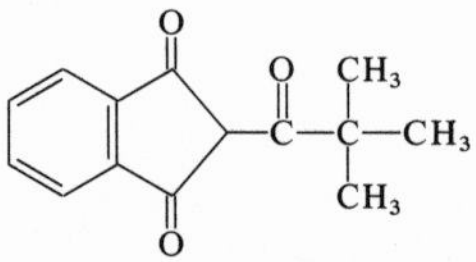 | Anticoagulant-type rodenticide; Norway rats, roof rats, mice | Motomco |
| **Plantvax***<br>F-461 | 2,3-Dihydro-5-carboxanilido-6-methyl-1,4-oxathiin-4,4-dioxide 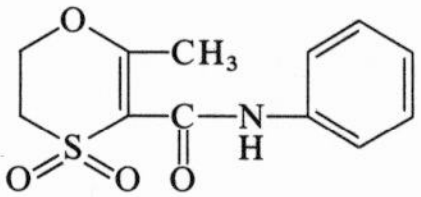 | Systemic fungicide against several pathogenic fungi, particularly *Basidiomycetes*, when applied to seed, soil, or foliage | Uniroyal |

| | | | |
|---|---|---|---|
| **PMA**<br>PMAS<br>Gallotox*<br>Phix*<br>Liquiphene* | Phenylmercuric acetate 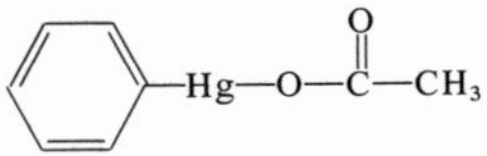 | Turf fungicide, paint preservative, and fungicide for sugarcane and rice seed; crabgrass control | Cleary<br>Guard |
| **Polyram*** | Mixture of 5.2 parts by weight (83.9%) of ammoniates of [ethylene-bis-(dithio-carbamate)]zinc with 1 part by weight (16.1%) ethylenebis[dithiocarbamic acid], bimolecular and trimolecular cyclic anhydrosulfides and disulfides | Foliage diseases on potatoes, apples, tomatoes, peanuts, sugar beets, other crops | FMC |
| **Preforan***<br>C-6989 | 2,4'-Dinitro-4-trifluoromethyldiphenyl-ether 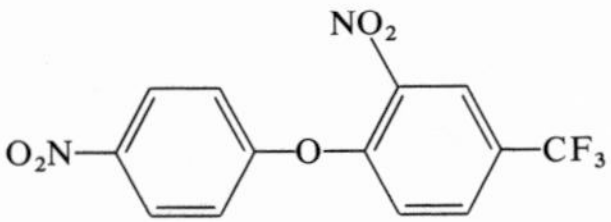 | Weed control in soybeans, field beans, snap beans, corn, rice, peanuts, okra, strawberries, Southern peas, English peas, small grains, cotton, and Brassica crops | Ciba |
| **Prometone**<br>Pramitol* | 2,4-bis(Isopropylamino)-6-methoxy-s-triazine 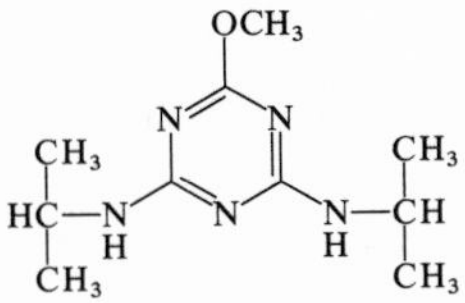 | Postemergence herbicide for nonselective or industrial weed control on noncrop land | Geigy |

| Pesticides common name and synonyms | Chemical name and formula | Major uses | Manufacturers |
| --- | --- | --- | --- |
| **Prometryne**<br>Caparol*<br>Primatol Q*<br>G-34161<br>Gesagard<br>Primaze<br>(Prometryne plus Atrazine) | 2,4-bis(Isopropylamino)-6-(methylthio)-s-triazine 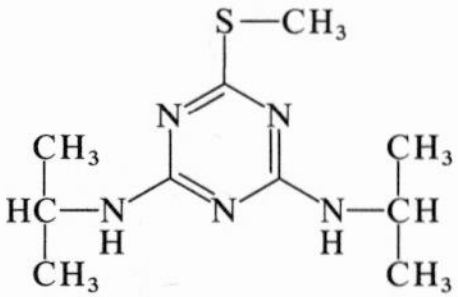 | Herbicide for the control of most annual grasses and broadleaved weeds in cotton and bluegrass grown for seed | Geigy |
| **Propachlor**<br>Ramrod* | 2-Chloro-N-isopropylacetanilide 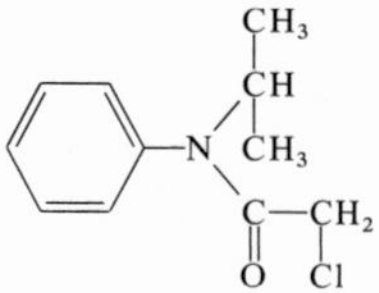 | Pre-emergence herbicide | Monsanto |
| **Propanil**<br>Rogue*<br>Surcopur*<br>Stam F-34*<br>Chem Rice* | 3,4-Dichloropropionanilide 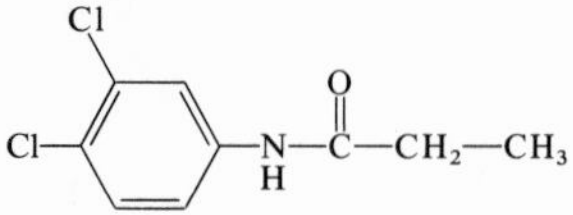 | Postemergence application to kill especially barnyard grass in rice | Monsanto<br>Chem. Ins.<br>Rohm & Haas |
| **Propargyl bromide** | 3-Bromo-1-propyne 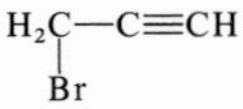 | In combination with methyl bromide and chloropicrin as soil fumigant | Dow |

| | | | |
|---|---|---|---|
| **Propazine**<br>Milogard*<br>Gesamil*<br>Primatol P*<br>G-30028 | 2-Chloro-4,6-bis(isopropylamino)-s-triazine 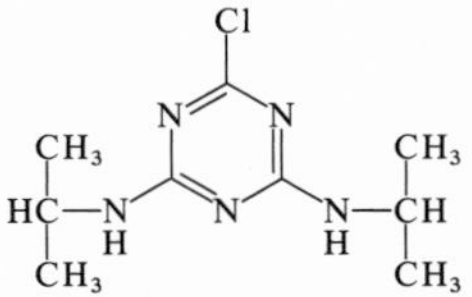 | Used as a pre-emergence herbicide for the control of most annual broad-leaved weeds and annual grasses in grain sorghum | Geigy |
| **Propham**<br>IPC<br>Chem Hoe* | Isopropyl N-phenylcarbamate 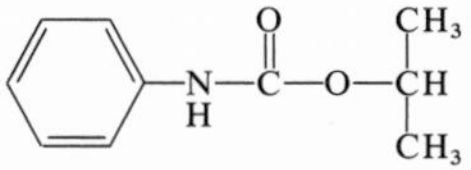 | Selective preplanting and pre-emergence herbicide; effective control of wild oats, many annual grasses and certain broad-leaved weeds | PPG |
| **Prophos**<br>Mocap*<br>V-C-9-104<br>ENT 27,318 | 0-Ethyl S, S-dipropyl phosphorodithioate 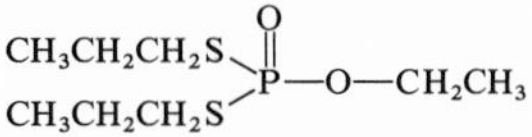 | Broad-spectrum activity as a soil nematocide-insecticide; nonfumigant, contact material with good soil movement and residual properties | Mobil |
| **Propyl isome**<br>Di-n-propylmaleate isosafrole condensate | Di-n-propyl 6,7-methylenedioxy-3-methyl-1,2,3,4-tetrahydronaphthalene-1,2-dicarboxylate 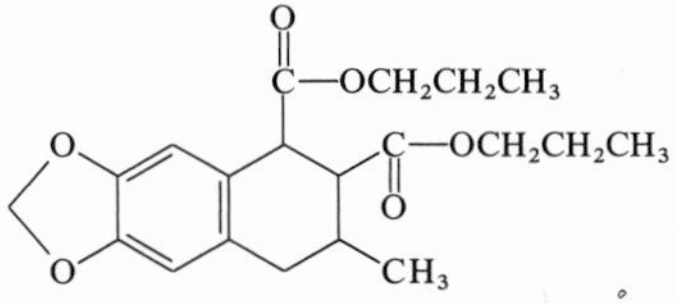 | Pyrethrin, allethrin, rotenone, and ryania synergist | Penick |
| **Pyrazon**<br>Pyramin* | 5-Amino-4-chloro-2-phenyl-3(2H)-pyridazinone 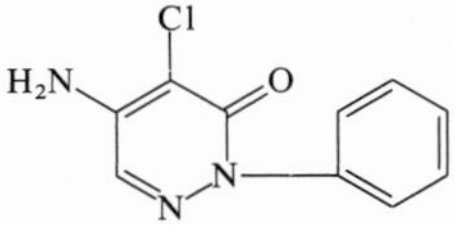 | Controls germinating broad-leaved weeds, including lamb's quarters | BASF<br>Amchem |

| Pesticides common name and synonyms | Chemical name and formula | Major uses | Manufacturers |
|---|---|---|---|
| **Pyrethrins**<br>Pyrethrum | Insecticidal components present in the flowers of *Pyrethrum cineraefolium*<br>R / R′<br>Pyrethrin I $CH_2\text{-}CH{=}CH\text{-}CH{=}CH_2$ $CH_3$<br>Cinerin I $CH_2\text{-}CH{=}CH\text{-}CH_3$ $CH_3$<br>Pyrethrin II $CH_2\text{-}CH{=}CH\text{-}CH{=}CH_2$ $COOCH_3$<br>Cinerin II $CH_2\text{-}CH{=}CH\text{-}CH_3$ $COOCH_3$<br>Jasmolin II $CH_2\text{-}CH{=}CH\text{-}CH_2\text{-}CH_3$ $COOCH_3$ 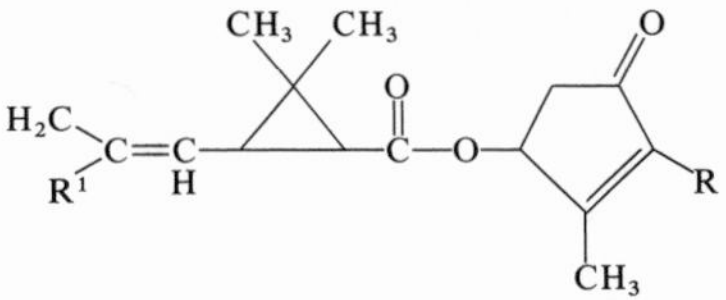<br>[Structure: $H_2C$, $R^1$, C=C(H), cyclopropane with $CH_3$, $CH_3$, C(=O)–O–, cyclopentenone with O, R, $CH_3$] | Knockdown and killing agent for household insects; protection of food in warehouses; dairy application and control of fruitfly on harvested fruits and vegetables; used with a synergist | Chem. Ins.<br>FMC<br>Penick<br>Prentiss<br>MGK |
| **8-Quinolinol**<br>Tumex*<br>Oxine<br>Chinosol*<br>Quinophenol<br>Bioquin | 8-Hydroxyquinoline sulfate 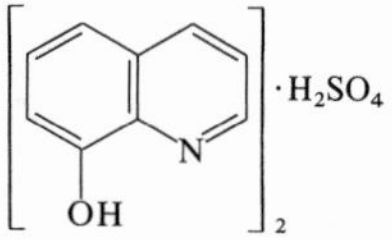<br>$[\text{8-hydroxyquinoline (N, OH)}]_2 \cdot H_2SO_4$ | Vegetable seedlings, citrus, grapes, and ornamentals; controls or prevents damping-off, vascular welts, seedling rots, blue mold, and other fungal diseases | Fisher<br>Merck |
| **Red squill** | Extract of the bulbs of *Urginea maritima* (Sea Onion) containing cardiac glycosides | Highly toxic to rats, less toxic to other animals | Chem. Ins.<br>Penick |

| | | | |
|---|---|---|---|
| **Ronnel**<br>Trolene*<br>Korlan*<br>Nankor*<br>Viozene*<br>Fenchlorfos | 0,0-Dimethyl-0-(2,4,5-trichlorophenyl) phosphorothioate 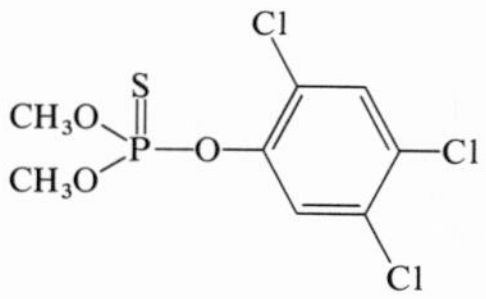 | Contact and systemic action controls flies and roaches, cattle and sheep ectoparasites | Dow |
| **Rotenone**<br>Protex | Rotenone 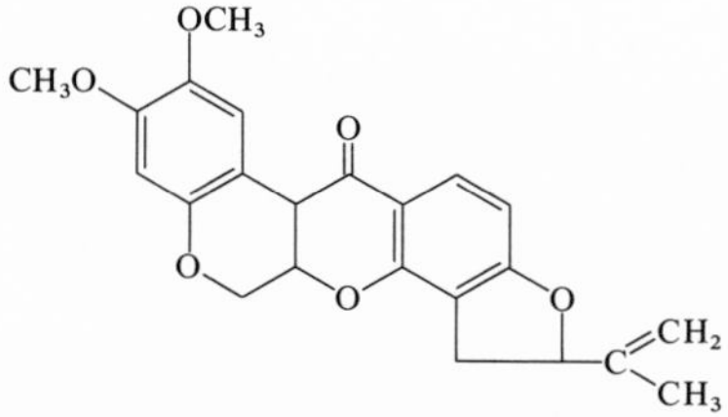 | Dusts for garden insects, lice, ticks on animals and cattle grubs; very toxic to fish, used for control and eradication of fish population | Chem. Ins.<br>FMC<br>Penick<br>Prentiss |
| **Ruelene*** | 4-tert-Butyl-2-chlorophenyl 0-methyl methylphosphoroamidate 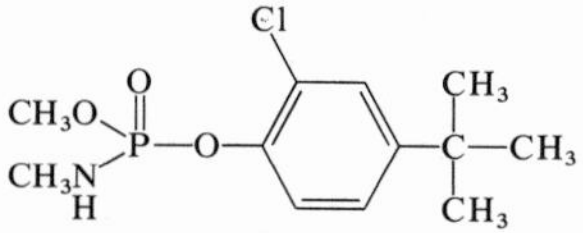 | Systemic insecticide for internal and external animal parasites | Dow |
| **Ryania**<br>Ryanicide*<br>Ryanexcel* | Ryanodine (extracted from stem wood of *Ryania speciosa*, a tropical shrub of Trinidad)<br>$C_{25}H_{35}NO_9$· or $C_{25}H_{37}NO_9$· | Insecticide for sugarcane and cornborers and larvae on apples | Penick |

| Pesticides common name and synonyms | Chemical name and formula | Major uses | Manufacturers |
|---|---|---|---|
| **Sabadilla**<br>Veratrine<br>Schoenocaulon<br>Ceradilla | Mixture of alkaloids known as veratrine, isolated from the seeds of *Schoenocaulum officinale* | Selective contact insecticide, effective against domestic pests and houseflies | Penick<br>Prentiss |
| **SBP-1382** | (5-Benzyl-3-furyl)methyl-2,2-dimethyl-3-(2-methylpropenyl)cyclopropanecarboxylate<br>$(H_3C)_2C{=}CH{-}C_3H_2(CH_3)_2{-}C(=O){-}O{-}CH_2{-}C_4H_2O{-}CH_2{-}C_6H_5$ | New household insecticide, effective without synergist | Penick |
| **Schradan**<br>OMPA | Octamethylpyrophosphoramide<br>$[(H_3C)_2N]_2P(=O){-}O{-}P(=O)[N(CH_3)_2]_2$ | Systemic insecticide for English walnuts, greenhouses, ornamentals | Pennsalt |
| **Semesan Bel***<br>Hydroxymercurinitrophenols | Hydroxymercurinitrophenol (12.5%) and hydroxymercurichlorophenol (3.8%)<br>$HO{-}C_6H_3(NO_2){-}Hg{-}OH$<br>$HO{-}C_6H_3(Cl){-}Hg{-}OH$ | Spray or dip treatment of Irish and sweet potatoes (seed pieces and slips) to protect against decay and control certain surface-borne diseases; preservation and control of large brown patch, dollar spot, copper spot and certain other diseases of fine turf | DuPont |

| | | | |
|---|---|---|---|
| **Sesone**<br>Crag Herbicide 1*<br>Disul-sodium<br>SES* | Sodium 2-(2,4-dichlorophenoxy)ethyl sulfate and sodium salt 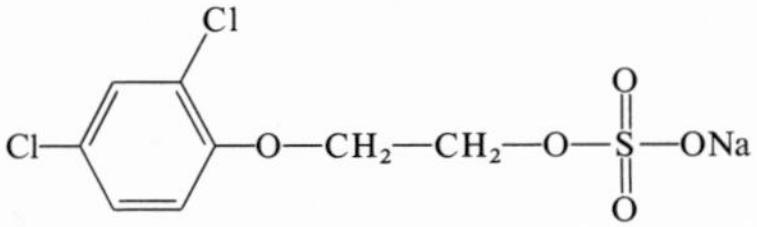 | Pre- and postemergence herbicide for strawberries, peanuts, potatoes, ornamentals | Amchem.<br>Carbide |
| **Siduron**<br>Tupersan* | 1-(2-Methylcyclohexyl)-3-phenylurea 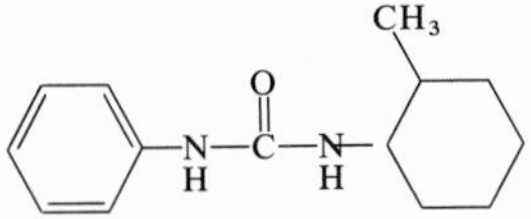 | Pre-emergence control of annual weed grasses such as crabgrass, foxtail, and barnyard grass, in newly seeded or established plantings of various grasses | DuPont |
| **Silica gel**<br>Dri-Die*<br>Silica aerogel | Treated silicon dioxide<br>$SiO_2$ | Household and grain storage insects | Grace |
| **Silvex**<br>Kuron*<br>Kurosal* | 2-(2,4,5-Trichlorophenoxy)propionic acid 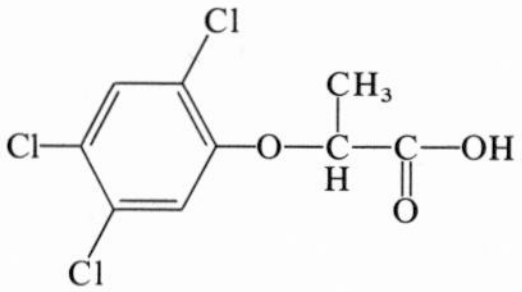 | Controls young broadleaved weeds, including chickweed, henbit, lamb's-quarters | Dow<br>Hercules |
| **Simazine**<br>Princep*<br>Gesapum*<br>Amizine*<br>Primatol S*<br>CDT<br>CET | 2-Chloro-4,6-bis(ethylamino)-s-triazine 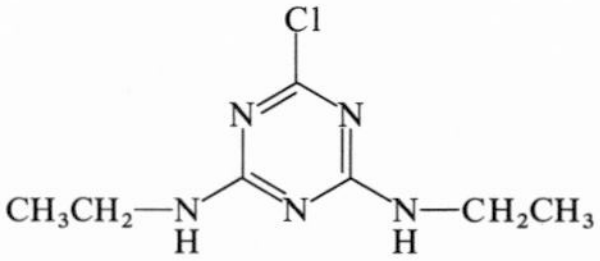 | Controls most annual grasses and broadleaved weeds in corn, sugarcane, fruits, nuts, asparagus, and turf | Geigy |

| Pesticides common name and synonyms | Chemical name and formula | Major uses | Manufacturers |
|---|---|---|---|
| **Sodium arsenite**<br>Sodium meta-arsenate<br>Chem-Sen | Sodium arsenite<br>$NaAsO_2$<br>plus<br>$As_2O_3$ | Potato vine killer; nonselective weed killer; aquatic herbicide; termites; debarking and destroying trees and tree stumps | Chem. Ins.<br>Chevron<br>Chipman<br>FMC<br>Pennsalt |
| **Sodium chlorate**<br>Altacide*<br>Chlorax*<br>Monoborochlorate<br>Drop-Leaf*<br>MBC*<br>Polybor chlorate*<br>Fall*<br>Tumbleaf*<br>Shed-A-Leaf "L" | Sodium chlorate<br>$NaClO_3$ | In the chlorine dioxide bleaching process for pulp and paper and in agriculture as a weed killer and cotton and soybean defoliant; also for metallurgical applications | American Potash<br>Chipman<br>Hooker<br>Pennsalt<br>PPG |
| **Sodium fluoride** | Sodium fluoride<br>NaF | Bait poison for ants and household insects | Allied<br>Alco<br>United |
| **Sodium fluoroacetate**<br>1080 | Sodium fluoroacetate<br>$F{-}CH_2{-}C(=O){-}ONa$ | Applied by pest control operators for rats, mice, ground squirrels and predators | Roberts<br>Tull |
| **Sok***<br>Banol* | 6(and 2)-Chloro-3,4-xylylmethyl-carbamate 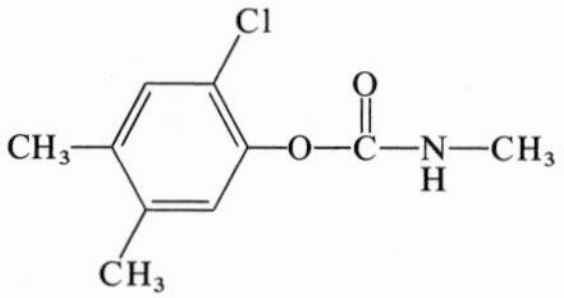<br>[structure: benzene ring with Cl, two $CH_3$ groups, and $-O-C(=O)-N(H)-CH_3$] | Control of ticks, mites, and insects on plants and animals | Tuco |

| | | | |
|---|---|---|---|
| **Solan** | 3′-Chloro-2-methyl-*p*-valerotoluidine 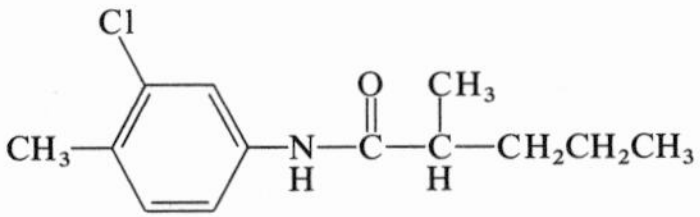 | Postemergence control of weeds in tomatoes | FMC |
| **Streptomycin**<br>Agrimycin*<br>Agri-Strep* | 2,4-Diguanidino-3,5,6-trihydroxycyclo-hexyl 5-deoxy-2-0-(2-deoxy-2-methyl-amino-α-glucopyranosyl)-3-formyl pentanofuranoside 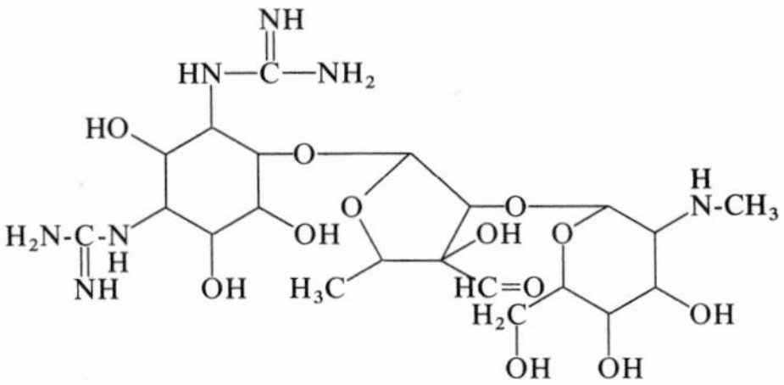 | Bactericide for fire blight on pears and apples | Merck<br>Pfizer |
| **Strobane***<br>Terpene polychlorinates | Terpene polychlorinates (65–66% Cl) consists of chlorinated camphene, penene, and related polychlorinates | Cotton insecticide | Tenneco |
| **Strychnine** | Alkaloid extracted from seeds of *Strychnos nux-vomica*, also strychnine sulfate<br>$C_{21}H_{22}N_2O_2$ | Control of predatory animals as well as field rodents such as gophers, pack rats, and field mice | (Imported) |
| **Succinic acid-dimethylhydrazide**<br>Alar*–85<br>B-nine | Succinic acid 2,2-dimethylhydrazide 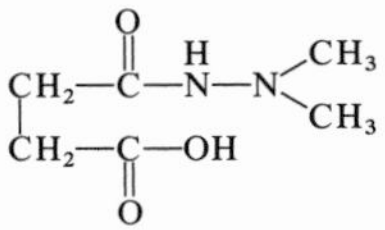 | Plant growth regulant, temporarily slows down vegetative growth, indirectly leads to flowering and fruiting of plants; used on apples and grapes | Uniroyal |

| Pesticides common name and synonyms | Chemical name and formula | Major uses | Manufacturers |
|---|---|---|---|
| **Sulfoxide**<br>Sulfox-Cide*<br>n-Octyl sulfoxide of isosafrole | 1-Methyl-2-(3,4-methylenedioxyphenyl) ethyl octyl sulfoxide<br>$CH_2$—$C(H)(CH_3)$—$S(=O)$—$(CH_2)_7CH_3$ | Pyrethrin and allethrin synergist; dairy, household and industrial insecticide, sprays, and aerosols | Penick |
| **Sulfur** | Sulfur<br>S | Fungicide for powdery mildews and other foliage diseases on fruit trees and vegetables; acaricide | Stauffer<br>FMC<br>Olin |
| **Sulfuryl fluoride**<br>Vikane* | Sulfuryl fluoride<br>$F$—$S(=O)_2$—$F$ | Structural fumigant for control of insect infestations | Dow |
| **Sulphenone** | *p*-Chlorophenyl phenyl sulfone<br>$C_6H_5$—$S(=O)_2$—$C_6H_4$—$Cl$ | Ovicide on apples, peaches, pears, almonds; poultry | Stauffer |
| **Sumithion***<br>Folithion*<br>Fenitrothion<br>Bay 41831 | O,O-Dimethyl O-(4-nitro-*m*-tolyl)phos-phorothioate<br>$(CH_3O)_2P(=S)$—O—$C_6H_3(CH_3)$—$NO_2$ | Contact insecticide effective against rice stem borers; selective acaricide, but of low ovicidal activity | Sumitomo<br>Chemagro |

| | | | |
|---|---|---|---|
| **2,4,5-T*** | 2,4,5-Trichlorophenoxyacetic acid 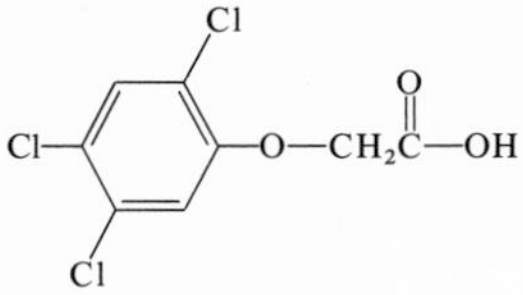 | Brush control on rangeland, pine tree stands, rights-of-way, aquatic weeds | Diamond<br>Dow<br>Hercules<br>Millmaster<br>Monsanto<br>Riverdale<br>Thompson-Hayward |
| **Tabutrex***<br>Tabatrex | Di-n-butyl succinate 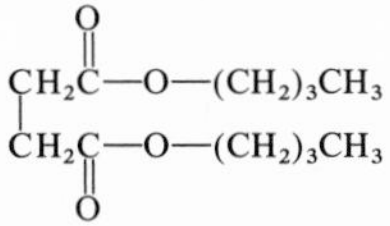 | Repellent for flies on farm animals and agricultural premises | Glenn |
| **Tandex*** | *m*-(3,3-Dimethylureido)phenyl tert-butylcarbamate 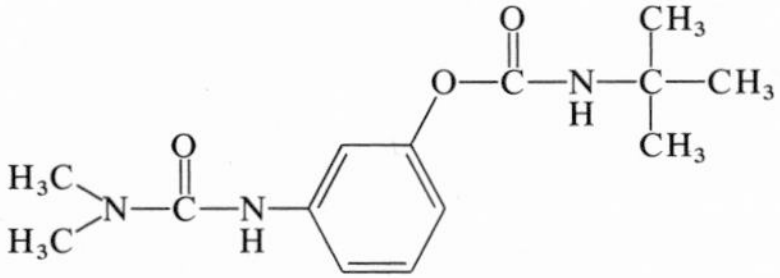 | Herbicide, soil sterilant for nonagricultural use | FMC |
| **2,3,6- TBA**<br>Trichlorobenzoic acid<br>HC-1281*<br>Fen-All*<br>Benzabor*<br>Trysben*<br>Ureabor* | Dimethylamine salt of 2,3,6-trichlorobenzoic acid and other trichlorinated benzoic acids 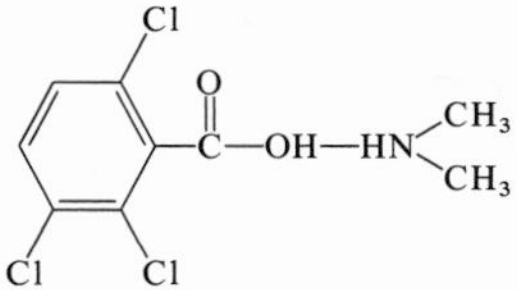 | Spray for nonselective control of certain undesirable broadleaved weeds and certain species of woody plants; intended for use on utility, highway, pipeline and railroad rights-of-way and other industrial sites | DuPont<br>Amchem<br>Tenneco |

| Pesticides common name and synonyms | Chemical name and formula | Major uses | Manufacturers |
|---|---|---|---|
| **TBP**<br>Tritac* | 1-[(2,3,6-Trichlorobenzyl)oxy]-2-propanol 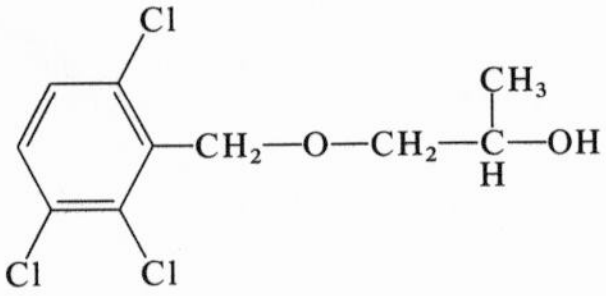<br>$Cl_3C_6H_2—CH_2—O—CH_2—CH(CH_3)—OH$ | Controls deep-rooted perennial broadleaved wheat; used only on noncrop land | Hooker |
| **TCA** | Trichloroacetic acid 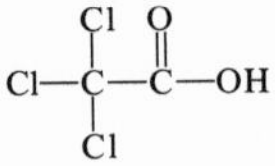<br>$Cl—C(Cl)_2—C(=O)—OH$ | Controls many germinating and established perennial grasses and broadleaved weeds | Dow<br>Monsanto |
| **TCBC**<br>Randox-T*<br>CDAA-T | Trichlorobenzylchloride 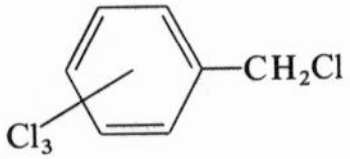<br>$Cl_3C_6H_2—CH_2Cl$ | Pre-emergence grass and broad-leaved weed selective herbicide; used only on noncrop land | Monsanto |
| **TCTP**<br>Tetrachlorothio-phene<br>Penphene* | Tetrachlorothiophene 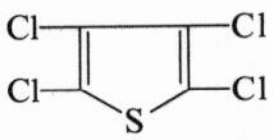<br>$C_4Cl_4S$ | Rootknot, meadow, stunt and dagger species of nematodes attacking tobacco | Pennsalt |
| **Temik***<br>UC21149 | 2-Methyl-2-(methylthio)propionaldehyde 0-(methylcarbamoyl)oxime 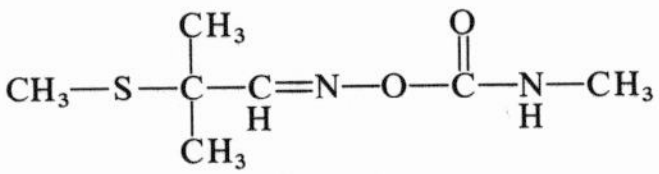<br>$CH_3—S—C(CH_3)_2—CH=N—O—C(=O)—NH—CH_3$ | To control aphids, mites, thrips, boll-weevils, flea beetles, wireworms, leaf miners, webworms, mealy bugs, leaf-hoppers, nematodes | Carbide |

| | | | |
|---|---|---|---|
| **TEPP**<br>Tetron*<br>Vapotone* | Tetraethyl pyrophosphate 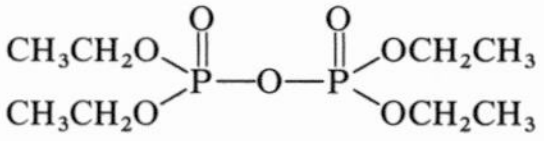 | Contact insecticide, very effective against active stages of mites and other softbodied insects | Amer. Potash<br>Chevron<br>Miller<br>Alco |
| **Terbacil**<br>Sinbar* | 3-tert-Butyl-5-chloro-6-methyluracil 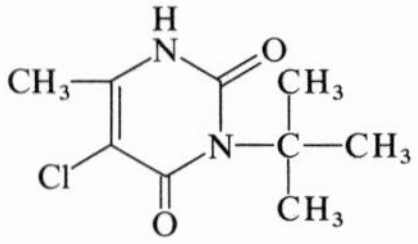 | Selective control of many annual and some perennial weeds in crops such as sugarcane, apples, peaches, citrus fruit and peppermint | DuPont |
| **Terbutol**<br>Azak* | 2,6-di-tert-Butyl-*p*-tolylmethylcarbamate 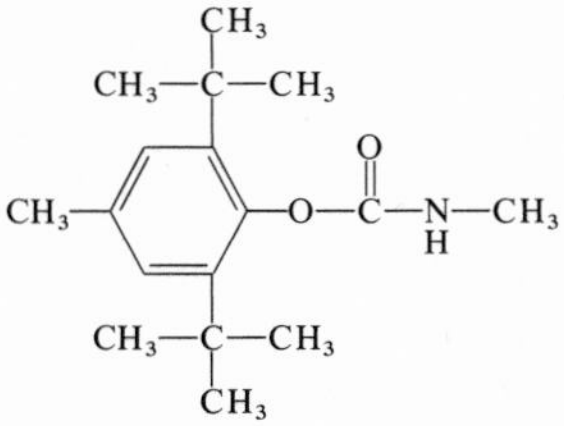 | Pre-emergence herbicide for control of crabgrass | Hercules |
| **Terrazole***<br>Terraclor*<br>Super X | 5-Ethoxy-3-trichloromethyl-1,2,4-thiadiazole 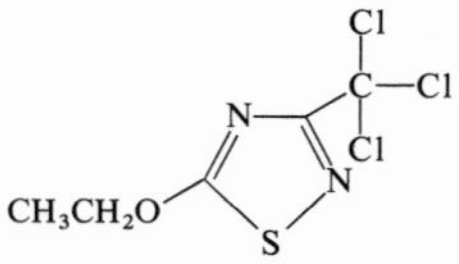 | Soil fungicide formulated with or without PCNB for use on cotton | Olin |

| Pesticides common name and synonyms | Chemical name and formula | Major uses | Manufacturers |
|---|---|---|---|
| **Tetrachloronitrobenzene** Fusarex* Folosan* TCNB Tecnazene | 1,2,4,5-Tetrachloro-3-nitrobenzene 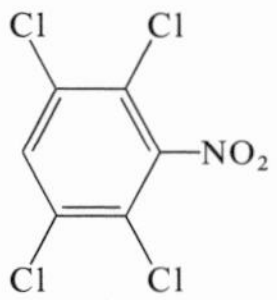 | Selective fungicide for dry rot of potato tubers | Bayer |
| **Tetradifon** Tedion* | 4-Chlorophenyl 2,4,5-trichlorophenyl sulfone 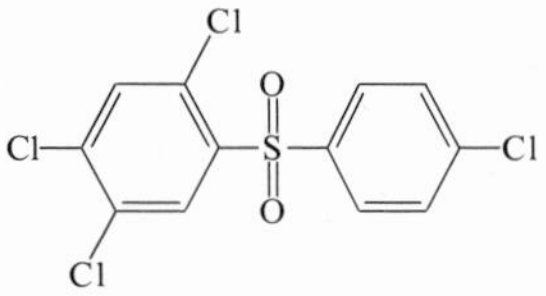 | Kills mites in all stages of development; deciduous fruits, citrus fruits, cotton and other crops | FMC |
| **Thallium sulfate** | Thallium sulfate $Tl_2SO_4$ | Ant, roach, mouse and rat control; cumulative poison | Asarco |
| **Thanite*** Isobornyl thiocyanoacetate | Isobornyl thiocyanoacetate (82%) and related compounds 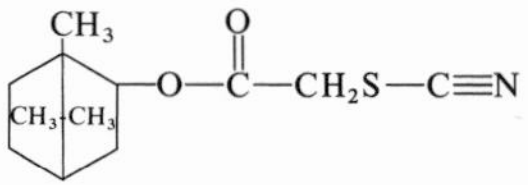 $-O-C(=O)-CH_2S-C\equiv N$ | Household insecticide with rapid knock-down and some residual action | Hercules |

| | | | |
|---|---|---|---|
| **Thiram**<br>Spotrete*<br>Arasan*<br>Delsan*<br>TMTDS<br>Pomarsol*<br>Vancide TM-95<br>Tersan*<br>Thylate*<br>Nomersan*<br>Mercuram* | bis(Dimethylthiocarbamoyl)disulfide<br>$(H_3C)_2N{-}C(=S){-}S{-}S{-}C(=S){-}N(CH_3)_2$ | Seed protectant, repellent for certain rodents and other animals, turf fungicide, and agricultural fungicide | DuPont<br>Cleary<br>Merck<br>Vanderbilt<br>Pennsalt<br>Vineland |
| **TIBA**<br>Regim-8*<br>Floraltone* | 2,3,5-Triiodobenzoic acid<br>$C_6H_2I_3{-}C(=O){-}OH$ | Soybean growth regulator and to promote flowering in apples | IMC<br>Amchem |
| **Toxaphene**<br>Alltex*<br>Toxakil* | Chlorinated camphene with 67–69% chlorine<br>$8(Cl){-}C_{10}H_{10}$ (bicyclic ring with $CH_2$ bridge, $=CH_2$, $CH_3$, $CH_3$) | Insecticide on cotton, livestock, grains, soybeans, forage crops, and vegetables | Hercules |
| **Triallate**<br>Avadex BW*<br>Far-Go* | S-2,3,3-Trichloroallyl-diisopropyl thiolcarbamate<br>$((H_3C)_2CH)_2N{-}C(=O){-}S{-}CH_2C(Cl){=}CCl_2$ | Pre-emergence application to control wild oats in barley, durum wheat; spring wheat, peas | Monsanto |

| Pesticides common name and synonyms | Chemical name and formula | Major uses | Manufacturers |
|---|---|---|---|
| **Tricamba**<br>Banvel T* | 3,5,6-Trichloro-*o*-anisic acid 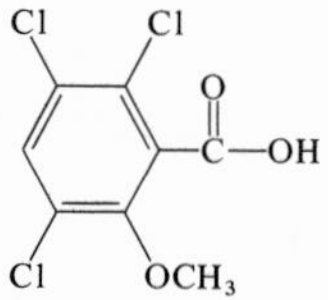 | Pre-emergence and foliar fungicide | Velsicol |
| **Trichlorfon**<br>Dylox*<br>Dipterex*<br>Neguvon*<br>Anthon*<br>Chlorphos | 0,0-Dimethyl(2,2,2-trichloro-1-hydroxyethyl)phosphonate 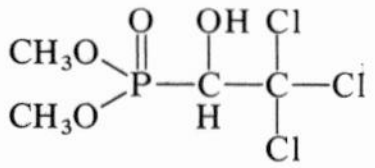 | Fly control; insects on crops; veterinary use for cattle and horse parasites | Chemagro |
| **Trifluralin**<br>Treflan* | α,α,α-Tifluoro-2,6-dinitro-N,N-dipropyl-*p*-toluidine 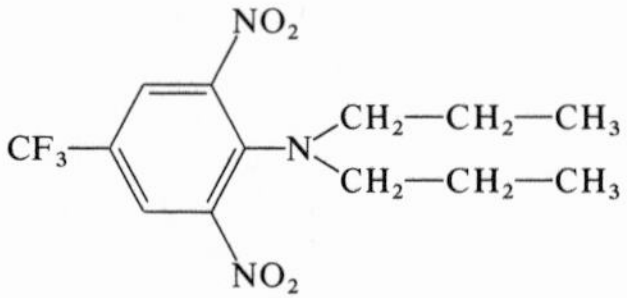 | Pre-emergence herbicide for use on cotton, soybeans, dry beans, sugarbeets, peas, and other vegetables | Elanco |
| **Triphenyltin hydroxide**<br>Du-Ter* | Triphenyltin hydroxide 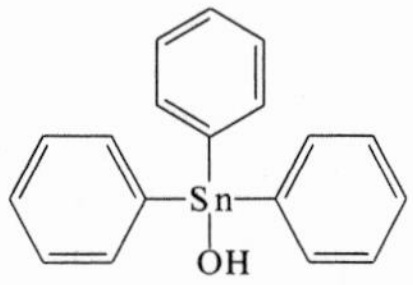 | Fungicide for control of potato blight; scab control on pecans | Thompson-Hayward |

| | | | |
|---|---|---|---|
| **Tropital*** | Piperonal bis[2-(2′-n-butoxyethoxy) ethyl]acetal 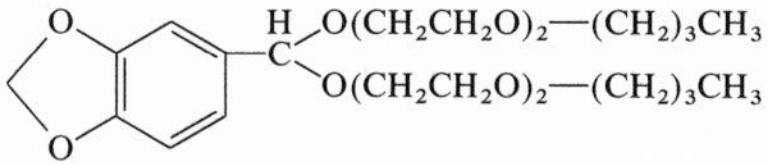 | Pyrethrin synergist for a wide variety of products, sprays and aerosols; helps provide "flushing" and quick knockdown in residual roach and ant spray | MGK |
| **Vernolate**<br>Vernam* | S-Propyl dipropylthiocarbamate 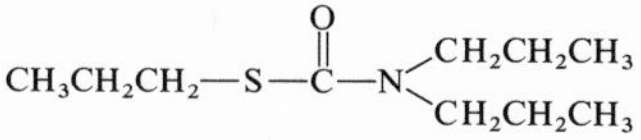<br>$CH_3CH_2CH_2{-}S{-}C(=O){-}N(CH_2CH_2CH_3)_2$ | Selective herbicide for control of annual grasses, many broadleaved weeds, as well as nutgrass in peanuts, soybeans and sweet potatoes | Stauffer |
| **Vitavax***<br>D-735 | 2,3-Dihydro-5-carboxanilido-6-methyl-1,4-oxathiin 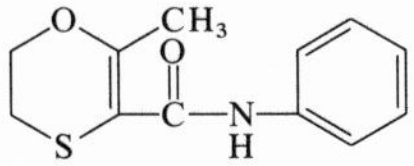 | Seed treatment fungicide for control of various smut and bunt diseases; particularly effective in controlling barley and wheat loose smut; systemic action | Uniroyal |
| **Vorlex***<br>Vorlex 201 | Composite of 20% methylisocyanate, chloropicrin and chlorinated $C_3$ hydrocarbons<br>$CH_3{-}N{=}C{=}S$ | Preplanting soil fumigant for weeds, fungi, nematodes, soil insects; for evergreens, vegetables, tobacco, cotton, small fruits, orchard trees | Morton |
| **Warfarin**<br>Coumafene<br>Dethmor* | 3-(α-Acetonylbenzyl)-4-hydroxycoumarin 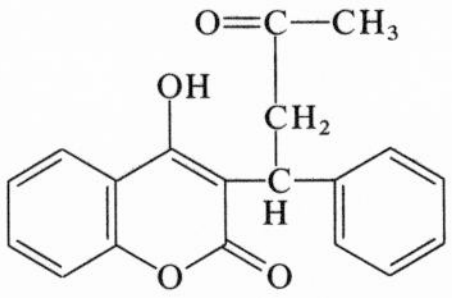 | Anticoagulant-type rodenticide effective against Norway rats, roof rats, mice, etc. | Penick<br>Prentiss |
| **Zinc phosphide** | Zinc phosphide<br>$Zn_3P_2$ | Used in baits for control of rats, mice, ground squirrels and prairie dogs | Hooker |

| Pesticides common name and synonyms | Chemical name and formula | Major uses | Manufacturers |
|---|---|---|---|
| **Zineb**<br>Parazate<br>Dithane Z-78*<br>Lonacol*<br>Amobam | Zinc ethylene bisdithiocarbamate<br>$\left[\begin{matrix} CH_2-NH-C(=S)-S^- \\ \vert \\ CH_2-NH-C(=S)-S^- \end{matrix}\right] Zn^{++}$ | To control many fungus diseases of vegetables, fruits and nuts, field crops and ornamentals | DuPont<br>Rohm & Haas<br>FMC<br>Pennsalt<br>Chem. Ins. |
| **Zinophos***<br>Cynem*<br>Nemafos* | 0,0-Diethyl-0-(2-pyrazinyl) phosphorothioate<br>$(CH_3CH_2O)_2P(=S)-O-C_4H_3N_2$ | For soil application on corn, vegetables, strawberries, sugar beets, cotton and peanuts | Cyanamid |
| **Ziram**<br>Zerlate*<br>Fuklasin<br>Pomarsol Z<br>Vancide MZ-96<br>Goodrite ZIP<br>ZIP*<br>Milbam | Zinc dimethyl dithiocarbamate<br>$(H_3C)_2N-C(=S)-S-Zn-S-C(=S)-N(CH_3)_2$ | For application as a spray or dust to control a wide range of fungus diseases of fruits, vegetables and ornamentals; also used as deer and rodent repellent as cyclohexyl-amine complex | DuPont<br>Vanderbilt<br>Pennsalt |

# Cross Index

Azofume see: Azobenzene

Bacillus Thuringiensis

Bakthane* see: Bacillus Thuringiensis

Balan* see: Benefin

Bandane*

Banol* see: Sok*

Bantrol* see: Ioxynil

Banvel* D see: Dicamba

Banvel* T see: Tricamba

Barban

Baron* see: Erbon

Barthrin

Basic cupric chloride see: Copper oxychloride sulfate

Basudin* see: Diazinon

Bay 2514* see: Dasanit*

Bay 29493* see: Fenthion

Bay 39007* see: Baygon*

Bay 41831 see: Sumithion*

Bayer 22555 see: Dexon*

Bayer 36205 see: Morestan*

Bayer 38819 see: Gophacide*

Bayer 44646 see: Matacil*

Baygon*

Bayluscide*

Baytex* see: Fenthion

Benefin

Benlate*

Bensulide

Benzabor* see: 2,3,6 TBA

Benzahex* see: BHC

Benzene hexachloride see: BHC; see also: Lindane

Benzex* see: BHC

Benzofume see: Azobenzene

4-Benzothienyl-N-methyl-carbamate see: Mobam*

(5-Benzyl-3-furyl)methyl-2,2-dimethyl-3-(2-methyl-propenyl)-cyclopropane-carboxylate see: SBP-1382

Betanal*

Betasan* see Bensulide

Bexide see: EXD

BHC

BHC, gamma isomer see: Lindane

Bidrin*

Binapacryl

B-nine see: Succinic acid-dimethylhydrazide

Binnell* see: Benefin

Bioquin* see: Copper-8-quinolinolate; see also: 8-Quinolinol

Biotrol BIB* see:

Bacillus Thuringiensis

Biphenyl

2,3,4,5-bis(2-Butylene)tetra-hydro-2-furaldehyde see: MGK Repellent 11

0,0-bis (*p*-Chlorophenyl)

acetimidoylphosphoramido-thioate see: Gophacide*

1,1-bis(*p*-Chlorophenyl -2,2,2-trichloroethanol see: Dicofol

bis(Dimethylthiocarbamoyl) disulfide see: Thiram

1,3-bis(1-Hydroxy-2,2,2-tri-chloroethyl)urea see: DCU

2,4-bis(Isopropylamino)-6-methoxy-s-triazine see: Prometone

2,4-bis(Isopropylamino)-6-(methylthio)-s-triazine see: Prometryne

2,2-bis(*p*-Methoxyphenyl-1,1,1-trichloroethane see: Methoxychlor

2,4-bis[(3-Methoxypropyl)-amino]-6-methylthio-s-triazine see: MPMT

Black Leaf 40* see: Nicotine

Black Leaf Products see: Nicotine

Bomyl*

Borascu* see: Borax*

Borax*

Bordeaux mixture

Boro-Spray* see: Borax*

Botran* see: DCNA

Bromacil see: Isocil

Brominal* see: Bromoxynil

5-Bromo-3-sec-butyl-6-methyluracil see: Isocil

3-(4-Bromo-3-chlorophenyl)-1-methoxy-1-methylurea see: Maloran*

Bromoethane see: Methyl bromide

Bromofume* see: EDP

3-(*p*-Bromophenyl)-1-methoxy-1-methylurea see: Metobromuron

Chem Rice* see: Propanil

Chem-Sen see: Sodium arsenite

Chinosol* see: 8-Quinolinol

Chlonitralid see: Bayluscide*

Chloramben see: Amiben*

Chloranil

Chlorasol* see:
Ethylene dichloride

Chlorax* see: Sodium chlorate

Chlorazine

Chlordane

Chlorfenson see: Ovex

Chlorinated camphene with 67–69% chlorine see: Toxaphene

Chlor Kil* see: Chlordane

Chlormequat see: Cycocel*

2-Chlorallyl diethyldithio-carbamate see: CDEC

Chlorobenzilate

2-Chloro-4,6-bis(diethylamino) s-triazine see: Chlorazine

2-Chloro-4,6-bis(ethylamino)-s-triazine see: Simazine

2-Chloro-4,6-bis(isopropyl-amino)-s-triazine see: Propazine

4-Chloro-2-butynyl-*m*-chloro-carbanilate see: Barban

0–[2-Chloro-1-(2,5-dichloro-phenyl)vinyl]0,0-diethyl phosphorothioate see:
Akton*

2-Chloro-2-diethylcarbamoyl-1-methylvinyl dimethyl phosphate see: Phosphamidon

2-Chloro-2′, 6′-diethyl-N-(methoxymethyl)acetanilide see: Lasso*

7-Chloro-4,6-dimethoxy-coumaran-3-one-2-spiro-1′ (2′-methoxy-6′-methylcyclo-hex-2′-en-4′-one) see: Griseofulvin

2-Chloro-4-ethylamino-6-isopropylamino-s-triazine see: Atrazine

2-Chloroethyltrimethyl ammonium chloride see: Cycocel*

Chlorofenson see: Ovex

Chloroform

Chloro IPC see: Chlorpropham

2-Chloro-N-isopropylacetan-ilide see: Propachlor

3′Chloro-2-methyl-*p*-valero-toluidine see: Solan*

0-2-Chloro-4-nitrophenyl 0,0-diethyl phosphorothioate see: Dicapthon

1-Chloro-2-nitropropane see: Lanstan*

Chlorophenothane see: DDT

*p*-Chlorophenoxyacetic acid see: 4-CPA

3-[*p*-(*p*-Chlorophenoxy) phenyl]-1,1-dimethylurea see: Chloroxuron

*p*-Chlorophenylbenzenesul-fonate see: Fenson*

*p*-Chlorophenyl-*p*-chloro-benzenesulfonate see: Ovex

3-(*p*-Chlorophenyl)-1,1-dimethylurea see: Monuron

3-(4-Chlorophenyl)-1,1-dimethylurea trichloroacetate see: Monoron TCA

*p*-Chlorophenyl phenyl sulfone see: Sulphenone

S-[*p*-Chlorophenylthio) methyl]0,0-diethyl phosphoro-dithioate see: Carbophenothion

S-([(p-Chlorophenyl)thio] methyl)0,0-dimethyl phos-phorodithioate see: Methyl trithion

4-Chlorophenyl 2,4,5-tri-chlorophenyl sulfone see: Tetradifon

Chloroneb

Chloropicrin

6-Chloropiperonyl chrysanthe-mumate see: Barthrin

6-Chloropiperonyl 2,2-di methyl-3-(2-methylpropenyl) cyclopropanecarboxylate see: Barthrin

Chloropropylate* see: Acaralate*

2-Chloro-1-(2,4,5-trichloro-phenyl)vinyl dimethyl phosphate see: Gardona*

Chloroxuron

6(and 2)-Chloro-3,4-xylyl-methylcarbamate see: Sok*

Chlorphenamidine

Chlorphos see: Trichlorfon

Chlorpropham

Cinerin see: Pyrethins

Ciodrin*

CIPC see: Chlorpropham

Clobber* see: Cypromid

CMA

Demeton-methyl see: Methyl-Demeton

Demeton-S-methyl see: Methyl-Demeton

Demosun* see: Chloroneb

2,4-DEP

Des-I-Cate see: Endothall

Dessin* see: Dinobuton

DET see: Diethyltoluamide

Dethmor* see: Warfarin

Dexon*

DHA see: Dehydroacetic acid

Diallate

N,N-Diallyl-2-chloroacetamide see: CDAA

S-(4,6-Diamino-s-triazin-2-ylmethyl)0,0-dimethyl phosphorodithioate see: Menazon

Diamond Arsonate Liquid see: MSMA

Diaphene* see: Bromsalans

Diazinon*

Dibasic lead arsenate see: Lead arsenate

Dibenzo 1,4-thiazine see: Phenothiazine

Dibrom* see: Naled

1,2-Dibromo-3-chloropropane see: DBCP

1,2-Dibromo-2,2-dichloroethyl dimethyl phosphate see: Naled

3,5-Dibromo-2-hydroxy-benzenanilide see: Bromsalans

3,5-Dibromo-4-hydroxybenzo-nitrile see: Bromoxynil

1,2-Dibromoethane see: EDP

Di-n-butyl succinate see: Tabutrex*

2,6-Di-tert-butyl-*p*-toly-methylcarbamate see: Terbutol

Dicamba

Dicapthon

Dichlobenil

Dichlone

Dichloral urea see: DCU

Dichloran see: DCNA

S-2,3-Dichloroallyl diiso-propylthiocarbamate see: Diallate

3,6-Dichloro-o-anisic acid see: Dicamba

2,6-Dichlorobenzonitrile see: Dichlobenil

1,1-Dichloro-2,2-bis(*p*-chlorophenyl)ethane see: DDD

1,1-Dichloro-2,2-bis(*p*-ethylphenyl)ethane see: Perthane*

2,4-Dichloro-6(o-chloro-aniline)-s-triazine see: Dyrene*

3'4'-Dichlorocyclopropane-carboxanilide see: Cypromid

1,4-Dichloro-2,5-dimethoxy-benzene see: Chloroneb

Dichloro-diphenyl-dichloro-ethane see: DDD

Dichloro-diphenyl-trichloro-ethane see: DDT

1,2-Dichloroethane see: Ethylene dichloride

4,4'-Dichloro-*a*-methylbenz-hydrol see: Dimite*

2,3-Dichloro-1,4-naphtho-quinone see: Dichlone

2,6-Dichloro-4-nitroaniline see: DCNA

1,1-Dichloro-1-nitroethane see: Ethide*

2',5-Dichloro-4'-nitrosalicyl-anilide ethanol amine see: Bayluscide*

2,4-Dichlorophenoxyacetic acid, amine salts and esters see: 2,4-D

4-(2,4-Dichlorophenoxy) butyric acid, salts, amine salts, and esters see: 2-4-DB

2-(2,4-Dichlorophenoxy)-propionic acid see: Dichlorprop

0-2,4-Dichlorophenyl 0,0-diethyl phosphorothioate see: Nemacide

3-(3,4-Dichlorophenyl)-1,1-dimethylurea see: Diuron

2,4-Dichlorophenyl ester of benzenesulfonic acid see: Genite (923)*

3-(3,4-Dichlorophenyl)-1-methoxy-1-methylurea see: Linuron

2,4-Dichlorophenyl *p*-nitro-phenyl ether see: Nitrofen

1,3-Dichloropropene, 3-3-dichloropropene, 1,2-dichloropropane, 2,3-dichloropropene and related $C_3$ chlorinated hydrocarbons, mixture see: D-D*

3,4-Dichloropropionanilide see: Propanil

2-Dimethylcarbamyl-3-methyl-5-pyrazolyl dimethylcarbamate see: Dimetilan

N,N-Dimethyl-2,2-diphenylacetamide see: Diphenamid

Dimethyldodecylamine acetate see: Penar*

Dimethyl 3-hydroxyglutaconate dimethyl phosphate see: Bomyl*

0,0-Dimethyl S-(N-methylcarbomylmethyl)phosphorodithioate see: Dimethoate

N,N-Dimethyl-N'-(2-methyl-4-chlorophenyl)-formamidine see: Chlorophenamidine

2,2-Dimethyl-3-(2-methylpropenyl)cyclopropanecarboxylic ester of N-(hydroxymethyl)-1-cyclohexane-1,2-dicarboximide see: Neo-pyamin*

0,0-Dimethyl-0-[4-(methylthio)-*m*-tolyl]-phosphorothioate see: Fenthion

0,0-Dimethyl 0-*p*-nitrophenyl phosphorothioate see: Methyl parathion*

0,0-Dimethyl 0-(4-nitro-*m*-tolyl)-phosphorothioate see: Sumithion*

0,0-Dimethyl-S-[4-oxo-1,2,3-benzotriazin-3(4H)-ylmethyl]-phosphorodithioate see: Azinphosmethyl

3-[2-(3,5-Dimethyl-2-oxocyclohexyl)-2-hydroxyethyl] glutarimide see: Cycloheximide

Dimethyl phosphate of *a*-methylbenzyl 3-hydroxy-*cis*-crotonate see: Ciodrin*

Dimethyl phosphate of 3-hydroxy-N,N-dimethyl-*cis*-crotonamide see: Bidrin*

Dimethyl phosphate of 3-hydroxy-N-methyl-*cis*-crotonamide see: Azodrin*

0,S-Dimethyl phosphoramidithioate see: Monitor*

0,0-Dimethyl phosphorodithioate of diethyl mercaptosuccinate see: Malathion

Dimethyl phthalate

Dimethyl-2,3,5,6-tetrachloroterephthalate see: DCPA

0,0-Dimethyl(2,2,2-trichloro-1-hydroxyethyl)phosphonate see Trichlorfon

Dimethyl(2,2,2-trichloro-1-hydroxyethyl)phosphorate ester of butyric acid see: Butonate

0,0-Dimethyl-0(2,4,5-trichlorophenyl) phosphorothioate see: Ronnel

1,1-Dimethyl-3(*a,a,a*-trifluoro-*m*-tolyl)urea see: Fluometuron

*m*-(3-3-Dimethylureido) phenyl tert-butylcarbamate see: Tandex*

Dimetilan

Dimite*

Dinex* see: Dinitrocyclohexylphonel

Dinitrocyclohexylphenol

4,6-Dinitrophenol-0-sec-butylphenol see: Dinoseb

2,4'-Dinitro-4-trifluoromethyl-diphenylether see: Preforan*

Dinobuton
Dinocap* see: Karathane*

Dinoseb

2,3-*p*-Dioxanedithiol-S,S-bis-(0,0-diethyl phosphorodithioate) see: Dioxathion

Dioxathion

Diphacin* see: Diphacinone

Diphacinone

Diphenamid

Diphenatrile

Diphenyl see: Biphenyl

Diphenylacetonitrile see: Diphenatrile

2-Diphenylacetyl-1,3-inandione see: Diphacinone

Diphenylamine

Diphenyl diimide see: Azobenzene

Dipropyl isoncinchomeronate see: MGK Repellent 326

Di-n-propylmaleate isosafrole condensate see: Propyl Isome

Di-n-propyl 6,7-methylenedioxy-3-methyl-1,2,3,4-tetrahydronaphthalene-1,2-dicarboxylate see: Propyl Isome

Dipterex* see: Trichlorfon

Diquat*

Disodium ethylenebisdithiocarbamate see: Nabam

Disodium methanearsonate see: DSMA

S-Ethyl diisobutylthiocarbamate see: Butylate

0-Ethyl S,S-dipropyl phosphorodithioate see: Prophos

S-Ethyl dipropylthiocarbamate see: EPTC

Ethylene

Ethylene dibromide see: EDB

Ethylene dichloride

Ethylenegycol bis(trichloroacetate) see: EGT

Ethylene oxide

S-Ethyl N-ethyl-N-cyclohexylthiocarbamate see: Cycloate

Ethyl formate

Ethyl guthion see: Azinphosethyl

S-Ethyl hexahydro-1H-azepine-1-carbothioate see: Molinate

Ethyl hexanediol

N-(2-Ethylhexyl)bicyclo-[2.2.1]-5-heptane-2,3-dicarboxamide see: MGK 264*

Ethyl hydrogen 1-propylphosphonate see: NIA 10637

Ethylmercury chloride

N-(Ethylmercury)-*p*-toluenesulfonanilide see: Ceresan M*

0-Ethyl-0-*p*-nitrophenyl phenylphosphonothioate see: EPN*

Ethyl Parathion see: Parathion*

0-Ethyl-S-phenylethylphosphonodithioate see: Dyfonate

0-[2-Ethylthio)ethyl]0,0-dimethyl phosphorothioate and S-[2-(Ethylthio)ethyl]0,0-dimethyl phosphorothioate see Methyl-Demeton

EXD

E-Z-off* see: Magnesium chlorate

E-Z off D* see: Def*

F-461 see: Plantvax*

Fall* see: Sodium chlorate

Falodin see: 2,4-DEP

Falone* see: 2,4-DEP

Far-Go* see: Triallate

Fenac*

Fen-All* see: 2,3,6 TBA

Fenitrothion see: Sumithion*

Fenson*

Fenchlorfos see: Ronnel

Fensulfothion see: Dasanit*

Fenthion

Fenuron

Fenuron TCA

Fenzaflor see: Lovozal

Ferbam

Fermate* see: Ferbam

Ferric dimethyl dithiocarbamate see: Ferbam

Flit MLO* see: Petroleum oils

Floraltone* see: TIBA

Fluometuron

Folex*

Folidol see: Parathion*

Folidol M* see: Methyl parathion*

Folithion* see: Sumithion*

Folosan* see: Tetrachloronitrobenzene

Folpet

Formaldehyde

Formalin see: Formaldehyde

Formetanate hydrochloride see: Carzol*

Forturf* see: Daconil 2787*

Fos-Fall "A" see: Def*

Fruitone see: 4-CPA

Fuklasin see: Ziram

Fumarin*

Fumazone* see: DBCP

Fundal see: Chlorphenamidine

Furadan* see: Carbofuran

Fusarex* see: Tetrachloronitrobenzene

F-W-925 see: Nitrofen

G-30028 see: Propazine

G-32293 see: Atratone

G-34161 see: Prometryne

Galecron* see: Chlorphenamidine

Gallotox* see: PMA

Gammexane* see: BHC

Gardona*

Genite (923)*

Genitol 923 see: Genite (923)*

Genitox* see: DDT

Isopropyl N-phenylcarbamate see: Propham

Isotox* see: BHC

Jasmolin II see: Pyrethrins

Karathane*

Karmex* see: Diuron

Kelthane see: Dicofol

Kemate see: Dyrene*

Kepone*

Kloben* see: Neburon

KOCN

Korlan* see: Ronnel

Kryocide* see: Cryolite

Kuron* see: Silvex

Kurosal* see: Silvex

Lambast* see: MPMT

Landrin*

Lannate* see: Methomyl

Lanstan*

Larvacide* see: Chloropicrin

Larvatrol* see: Bacillus Thuringiensis

Lasso*

Lead arsenate

Lebaycid see: Fenthion

Lenacil

Lethane 384

Lime sulfur

Limit* see: CDAA

Lindane

Lintox see: BHC

Linuron

Liquiphene see: PMA

LM seed protectant see: Ortho LM

Lonacol* see: Zineb

Lorox* see: Linuron

Lovozal*

MAA see: DSMA

Mafu* see: Dichlorvos

Magnesium chlorate

Magron* see: Magnesium chlorate

Malachite green see: Copper carbonate, basic

Malaspray* see: Malathion

Malathion

Maleic hydrazide see: MH

Maloran*

MAMA

Maneb

Manganese dimethyl dithiocarbamate and mercaptobenzothiazole see: Niacide*

Manganous ethylenebisdithiocarbamate see: Maneb

Manzate* see: Maneb

Marlate* see: Methoxychlor

Matacil*

MBC* see: Sodium Chlorate

MCA 600* see: Mobam*

MCP see: MCPA

MCPA

MCPB

MCPP

MEB see: Maneb

Mecarbam

Mecopar see: MCPP

Mecopex* see: MCPP

Mecoprop see: MCPP

Memmi*

Menazon

Mephanac* see: MCPA

Mercaptodimethur see: Mesurol*

N-(Mercaptomethyl)phthalimide S-(O,O-dimethyl phosphorodithioate) see: Imidan*

Mercaptophos see: Demeton

Mercuram* see: Thiram

Mercuric chloride

Merphos* see: Folex*

Merthiolate* see: Elcide *73

Mesurol*

Metaldehyde

Metam see: Metham

Meta-Systox* see: Methyl-Demeton

Metham

Methar see: DSMA

Methomyl

Methoxone* see: MCPA

Methoxychlor

Methoxy DDT see: Methoxychlor

Methyl bromide

Methyl 1-(butylcarbamoyl)-2-benzimidazole carbamate see: Benlate*

*m*(1-Methylbutyl)phenyl methylcarbamate (75%) and

*m*(1-ethylpropyl)phenyl methylcarbamate (25%), mixture see: Bux Ten*

2-Methyl-4-chlorophenoxy-acetic acid see: MCPA

4-(2-Methyl-4-chlorophenoxy) butyric acid see: MCPB

2-(2-Methyl-4-chlorophenoxy) propionic acid see: MCPP

1-(2-Methylcyclohexyl)-3-phenylurea see: Siduron

Methyl-Demeton

2-Methyl-4,6-dinitrophenol sodium salt see: DNOC

2-(1-Methylheptyl)-4,6-dinitrophenyl crotonate see: Karathane*

Methyl *m*-hyroxycarbamanilate *m*-methylcarbanilate see: Betanal*

Methylisocyanate, chloropicrin and chlorinated $C_3$ hydro-carbons see: Vorlex*

Methyl mercury dicyandiamide see: Panogen*

Methyl mercury 2,3-dihydroxy-propyl mercaptide and methyl mercury acetate mixture see: Ceresan L*

Methyl mercury 8-hydroxy-quinolate see: Ortho LM

S-Methyl N-[(methylcarba-moyl)oxy] thioacetimidate see: Methomyl

1-Methyl-2-(3,4-methylene-dioxyphenyl)ethyl octyl sulfoxide see: Sulfoxide

2-Methyl-2-(methylthio) propionaldehyde O-(methyl-carbamoyl)oxime see: Temik*

Methyl parathion*

3-(2-Methylpiperidino)propyl 3,4-dichlorobenzoate see: Pipron

3-(1-Methyl-2-pyrrolidyl) pyridine see: Nicotine

6-Methyl-2,3-quinoxaline-dithiol cyclic-S,S-dithio-carbonate see: Morestan*

4-(Methylsulfonyl)-2,6-dinitro-N,N-dipropylaniline see: Nitralin

4-(Methylthio)3-5-xylyl-methyl-carbamate see: Mesurol*

Methyl trithion

Meticide* see: Methyl parathion

Metmercapturon see: Mesurol*

Metobromuron

Metron* see: Methyl parathion*

Mevinphos

MGK 264*

MGK Repellent 11

MGK Repellent 326

MGK Repellent 874

MH

MH-30* see: MH

Mico-Fume* see: Dazomet

Milbam see: Ziram

Mildex* see: Karathane*

Miller 531* see: Cadmium-calcium-copper-zinc-chromate complex

Miller 658 see: Copper zinc chromate

Milogard* see: Propazine

Mirex

MnEBD see: Maneb

Mobam*

Mocap* see: Prophos

Molinate

Monitor*

Monoammonium methyl-arsenate see: MAMA

Monoborochlorate see: Sodium chlorate

Monosodium acid methane-arsonate see: MSMA

Monuron

Monuron TCA

Morestan*

Morkit* see: Anthraquinone

Morocide* see: Binapacryl

Morsodren* see: Panogen*

MPMT

MSMA

Murfotox* see: Mecarbam

Murotox* see: Mecarbam

Murvesco* see: Fenson*

Muscatox* see: Coumaphos

Mylone* see: Dazomet

N-2790 see: Dyfonate

Nabam

Naled

Nankor* see: Ronnel

Naphthalene

Naphthalene acetamide

Naphthaleneacetic acid

Naphthenic acids, copper salts see: Copper naphthenate

1-Naphthyl N-methylcarbamate see: Carbaryl

N-1-Naphthylphthalamic acid see: Naptalam

a-Naphthylthiourea see: Antu*

Naptalam

Navadel* see: Dioxathion

NC 5016 see: Lovozal*

Neburon

Neguvon* see: Trichlorfon

Nemacide

Nemafos* see: Zinophos*

Nemagon* see: DBCP

Nemax* see: Chloropicrin

Nemex* see: D-D*

Neobor* see: Borax*

Neo-pynamin*

NIA 5961 see: Lanstan*

NIA 10637

Niacide*

Niagaramite* see: Aramite*

Nialate* see: Ethion

Nickel sulfate and manganous ethylenbis[dithiocarbamate] see: Dithane S-31*

Nicotine

Nicotine sulfate see: Nicotine

NIP see: Nitrofen

Niran* see: Parathion*

Nitralin

2-Nitro-1,1-bis(p-chlorophenyl) butane and 2-nitro-1,1-bis(p-chlorophenyl) propane see: Dilan*

Nitrofen

No Bunt 40 or 80* see: HCB

Nomersan* see: Thiram

Norbormide

Norea

Novege* see: Erbon

NPA see: Naptalam

NPD* see: Aspon*

Octachloro-4-7-methanotetrahydroindane (60%) and related compounds (40%), mixture see: Chlordane

Octacide 264 see: MGK 264*

Octalox* see: Dieldrin

Octamethylpyrophosphoramide see: Schradan

N-Octylbicycloheptenedicarboximide see: MGK 264*

n-Octyl sulfoxide of isosafrole see: Sulfoxide

Oko* see: Dichlorvos

OMPA see: Schradan

OPP see: Orthophenylphenol

Orchex* see: Petroleum oils

Ordram* see: Molinate

Ortho 5353 see: Bux Ten*

Ortho 9006 see: Monitor*

Orthocide* see: Captan*

Orthodichlorobenzene

Ortho Dual Paraquat see: Paraquat

Ortho-Klor* see: Chlordane

Ortho LM

Ortho MC* see: Magnesium chlorate

Ortho Paraquat Cl see: Paraquat

Orthophenylphenol

Orthophos see: Parathion*

Ortho Phosphate Defoliant see: Def*

Orthoxenol see: Orthophenylphenol

Ovex

Ovotran* see: Ovex

7-Oxabicyclo(2.2.1) heptane-2,3-dicarboxylic acid see: Endothall

Oxine see: 8-Quinolinol

Oxirane see: Ethylene oxide

Oxythionquinox see: Morestan*

Pandrinox* see: Panogen*

Panodrin A13* see: Panogen*

Panogen*

Panoram D-31* see: Dieldrin

Panterra* see: Panogen*

Panthion* see: Parathion*

Paradichlorobenzene

Paraquat

Parathion*

Parawet* see: Parathion*

Parazate see: Zineb

Paris green

PAS

Patoran* see: Metobromuron

Propoxur see: Baygon*

S-Propyl butylethylthio-carbamate see: Pebulate

S-Propyl dipropylthiocarba-mate see: Vernolate

Propyl Isome

Protex see: Rotenone

Prussic acid see: Hydrogen cyanide

Pyramin* see: Pyrazon

Pyrazon

Pyrethrins

Pyrethrum see: Pyrethrins

Puratized* agricultural spray see: PAS

Queletox* see: Fenthion

8-Quinolinol

Quinomethionate see: Morestan*

Quinophenol see: 8-Quinolinol

2,3-Quinoxalinedithiol cyclic trithiocarbamate see: Eradex

R-2063 see: Cycloate

R-4461 see: Bensulide

R-4572 see: Molinate

Rabon* see: Gardona*

Rack Granular* see: Atrazine

Radapon* see: Dalapon

Rad-E-Cate 35* see: Cacodylic acid

Ramrod* see: Propachlor

Randox* see: CDAA

Randox-T* see: TCBC

Raticate* see: Norbormide

RCI 49-155 see: Orthophenylphenol

RCI 49-162 see: PCP

Red squill

Regim-8* see: TIBA

Reglone* see: Diquat*

Resitox* see: Coumaphos

Retard* see: MH

Rhothane* see: DDD

Rogor* see: Dimethoate

Rogue* see: Propanil

Ro-neet* see: Cycloate

Ronnel

Rootone* see: Naphthalene acetamide

Rotenone

Ruelene*

Ryanexcel* see: Ryania

Ryania

Ryanicide* see: Ryania

Ryanodine see: Ryania

S-6000 see: Cypromid

Sabadilla

Sarolex* see: Diazinon*

SBP-1382

Schoenocaulon see: Sabadilla

Schradan

SD3562 see: Bidrin*

SD9098 see: Akton*

SD9129 see: Azodrin*

Seedrin see: Aldrin

Semasan* see: Hydroxymercurichlorophenol

Semesan, Bel*

SES* see: Sesone

Sesone

Sevin* see: Carbaryl

Shed-A-Leaf "L" see: Sodium chlorate

Shoxin* see: Norbormide

Sinbar see: Terbacil

Siduron

Silica aerogel see: Silica gel

Silica gel

Silicon dioxide, treated see: Silica gel

Silvex

Simazine

Sinox see: DNOC

Slo-Gro* see: MH

SMDC see: Metham

Snip* Fly Bands see: Dimetilan

Sodium arsenite

Sodium chlorate

Sodium 2-(2,4-dichloro-phenoxy)ethyl sulfate and sodium salt see: Sesone

Sodium ethylmercurithiosalicy-late see: Elcide *73

Sodium fluoaluminate see: Cryolite

Sodium fluoride

Sodium fluoroacetate

Sodium metaarsenite see: Sodium arsenite

Sodium methyldithiocarbamate see: Metham

Sodium pentachlorophenate see: PCP

Thuricide* see: Bacillus Thuringiensis

Thylate* see: Thiram

TIBA

Tiguvon* see: Fenthion

Tillam* see: Pebulate

TMTDS see: Thiram

TOK E-25 see: Nitrofen

Topane* see: Orthophenylpheno

Tordon* see: Picloram

Toxaphene

Toxakil* see: Toxaphene

Treflan* see: Trifluralin

Triallate

Triazine* see: Dyrene*

Tri-ban* see: Pival*

Tributyl 2,4-dichlorobenzyl-phosphonium chloride see: Phosfon*

S,S,S-Tributyl phosphoro-trithioate see: Def*

S,S,S-Tributyl phosphoro-trithioite see: Folex*

Tricamba

Trichloracetic acid see: TCA

Trichlorfon

S-2,3,3-Trichloroallyl diisopropylthiolcarbamate see: Triallate

3,5,6-Trichloro-o-anisic acid see: Tricamba

Trichlorobenzylchloride see: TCBC

Trichlorobenzoic acid see: 2,3,6 TBA

2,3,6-Trichlorobenzyloxypro-panol see: TBP

1,1,1-Trichloro-2,2-bis (*p*-chlorophenyl)ethane see: DDT

N-Trichloromethylthio-4-cyclo-hexene 1,2-dicarboximide see: Captan*

N-(Trichloromethylthio)-phthalimide see: Folpet

Trichloronitromethane see: Chloropicrin

2-(2,4,5-Trichlorophenoxy) ethyl 2,2-dichloropropionate see: Erbon

2,4,5-Trichlorophenoxyacetic acid see: 2,4,5-T

2-(2,4,5-trichlorophenoxy)-propionic acid see: Silvex

2,3,6-Trichlorophenylacetic acid or sodium salt see: Fenac*

Tridex see: EXD

Tri-Fene* see: Fenac*

Trifenson* see: Fenson*

*a,a,a,*-Trifluoro-2,6-dinitro-N,N-dipropyl-*p*-toluidine see: Trifluralin

Trifluralin

2,3,5-Triiodobenzoic acid see: TIBA

2,3,5-Trimethylphenyl methyl-carbamate 3,4,5-Trimethyl-phenyl methyl carbamate see: Landrin

Triphenyltin hydroxide

Tris [(2,4-dichlorophenoxy)-ethyl] phosphite see: 2,4-DEP

Tritac* see: TBP

Trithion* see: Carbophenothion

Trolene* see: Ronnel

Trona see: Borax*

Tronabor* see: Borax*

Tropital*

Tropotox* see: MCPB

Tumbleaf* see: Sodium chlorate

Tumex* see: 8-Quinolinol

Tupersan* see: Siduron

UC21149 see: Temik*

Unden* see: Baygon*

Urab* see: Fenuron TCA

Ureabor* see: 2,3,6 TBA

*Urginea maritima* extract see: Red squill

Vancide FE-95 see: Ferbam

Vancide MZ-96 see: Ziram

Vancide TM-95 see: Thiram

Vapam* see: Metham

Vapotone* see: TEPP

Vapona* see: Dichlorvos

V-C-9-104 see: Prophos

V-C-13* see: Nemacide

Vegadex* see: CDEC

Venzar* see: Lenacil

Veratrine see: Sabadilla

Verbigen see: Amiben*

Vernam* see: Vernolate

Vernolate

Verton* D see: 2,4-D

Vidden D see: D-D*

Vikane* see: Sulfuryl fluoride

Viozene* see: Ronnel

## 2 Literature Review on Mutagenicity of Pesticides

This review was compiled from a manual search of the literature, from the *Environmental Mutagen Information Center (EMIC)* registry, and from *MEDLARS*.

**Literature Review on Mutagenicity of Pesticides** References are given in full in the Bibliography.

| Pesticide | Organism in Which Tested | Assay System | DOSE Range | DOSE Minimum Effective Dose | Biological Effect |
|---|---|---|---|---|---|
| Acrylonitrile | Yeast[298] | t-RNA | | | Cynoethylation |
| Acrylonitrile | Yeast[398] | t-RNA | | | Pseudo-uridine modification |
| Atrazine | Barley[395] | Anther | 1000 ppm-soaked | | Slight effect on meiosis ($C_1$)<br>Slight effect on meiosis ($C_2$) |
| Captan | Human embryo cells[222] | L-132 cells | 10 mcg/ml | | Inhibition of DNA |
| Captan | Mice[119] | Sperm | 9 mg/kg<br>500 mg/kg | | No increase in frequency of dominant lethals |
| Captan | Rat kangaroo[222] | Somatic and germ cells | 1.25 to 5.0 mcg/ml | | Chromosome aberrations |
| Carbaryl | Barley[395] | Anther | 1000 ppm-soaked<br>500 ppm-sprayed | | No effect on meiosis ($C_1$)<br>Abnormal meiosis ($C_2$) |
| Carbaryl | Plant[14] | Root tips | 0.5 and 0.25 saturated | | Abnormal mitosis<br>Chromosome aberrations |

| Pesticide | Organism in Which Tested | Assay System | DOSE Range | DOSE Minimum Effective Dose | Biological Effect |
|---|---|---|---|---|---|
| Chloroform | *Allium cepa*[352] | Root tips | Saturated to 0.005% | 0.025% | C-mitosis chromosome aberrations |
| Chloroform | *Allium cepa*[283] | Root tips | | | C-mitosis |
| Chlorprophan | *Allium cepa*[238] | Plant cells | 2.5, 5, 10, 20 40, 80 ppm | | C-mitotic effect Nuclear disintegration |
| 2,4–D | Narcissus[320] | Root tips | 0.01, 0.05, 0.1% | | C-mitosis Chromosome aberrations |
| 2,4–D | Cotton (Acla 44)[32] | Cotyledons | $10^{-3}$ M–$10^{-4}$ M | | Effects nucleic acid synthesis |
| 2,4–D | *Allium cepa*[320] | Root tips | 0.01, 0.05, 0.1% | | C-mitosis Chromosome aberrations |
| 2,4–D | *Vicia faba*[325] | Root tips | 0.001% to 1.0% | 0.001% | Abnormal mitosis |
| 2,4–D | *Allium cepa*[98] | Root tips | 25 to 500 ppm | 25 ppm | Chromosome aberrations |
| 2,4–D | *Tradescantia*[325] | | 0.001% to 1.0% | 0.001% | Abnormal mitosis |
| DCNA | Barley[395] | Anther | 1000 ppm-soaked | | Slight effect on meiosis ($C_1$) |
| | | | 500 ppm-sprayed | | High abnormal meiosis ($C_2$) |

| | | | | | |
|---|---|---|---|---|---|
| DDT | Mice[119] | Sperm | 105 mg/kg | | No increase in frequency of dominant lethals |
| DDT | *Allium cepa*[373] | Root tips | Saturated solution | | C-mitosis and chromosome breaks |
| DDT | *Trigonella foreum graecum*[373] | Root tips | Saturated solutions | | C-mitosis and chromosome breaks |
| DDT | rat* | sperm | 50–70 mg/kg | 50 mg/kg | increase in frequency of dominant lethals |
| DDT | marsupial celline* | somatic cell | 10–50 ppm | 10 ppm | chromosome breaks and exchange figures |
| Dichlorvos | Onion[327] | Root tips | 0.5 sq cm to 6.0 sq cm | | Chromosome breaks |
| Dicomba | Barley[395] | Anther | 1000 ppm-soaked | | Abnormal meiosis ($C_1$) |
| | | | 500 ppm-sprayed | | Abnormal meiosis ($C_2$) |
| Dieldrin | *Crepis capillaris*[240] | Sprouts | 10% solution | | C-mitotis effect, no chromosome breaks observed |
| Endothall | *Crepis capillaris*[389] | Plant cells | | | Chromosome aberrations |

*Palmer, K., Green, S., and Legator, M.: unpublished data (1971).

| Pesticide | Organism in Which Tested | Assay System | DOSE Range | Minimum Effective Dose | Biological Effect |
|---|---|---|---|---|---|
| Endrin | Barley[395] | Anther | 1000 ppm-soaked | | No effect on meiosis ($C_1$) |
| | | | 500 ppm-sprayed | | No effect on meiosis ($C_2$) |
| Ethylene Oxide | Fungi[386] | 0.025 M | | | Point mutations and reverse mutations |
| Ethylene Oxide | *Neurospora crassa*[198] | Conidia | 0.14 M | | Point mutations and reverse mutations |
| Ethylene Oxide | Maize[121] | Plant cells | 1 part E.O. to 20 parts air | | Chromosome breaks |
| Ethylmercury chloride | *Triticum*[210] | Root tips | 0.5 to 1% | | Mitotic aberrations |
| Ethylmercury chloride | *Secale cereale*[210] | Root tips | 0.5 to 1% | | Mitotic aberrations |
| Ferbam | *Aspergillus niger*[289] | Spores | 1000 ppm | | Morphological mutants and reverse mutations |
| Ferbam | *Allium cepa*[289] | Root tips | 240 ppm | | Chromosome aberrations |
| Formaldehyde | *Drosophila melanogaster*[347] | Sperm | 0.033–0.50 M | | Induced very low incidence of recessive lethals but enhanced the effect of x-rays |

| | | | | | |
|---|---|---|---|---|---|
| HCN | Mammalian embryo[85] | Heart, spleen, and liver cells | $0.20 \times 10^{-2}$ to $0.88 \times 10^{-6}$ | | Induced some nuclear abnormalities |
| HCN | *Vicia faba*[226] | Root tips | $4 \times 10^{-4}$ M | | Chromosome breaks |
| Isocil | Barley[395] | Anther | 1000 ppm-soaked | | Abnormal meiosis ($C_1$) |
| | | | 500 ppm-sprayed | | Abnormal meiosis ($C_2$) |
| KCN | *Vicia faba*[197] | Root tips | 400 $\mu$M/L<br>1000 $\mu$M/L | | Chromosome breaks |
| Lindane | Onion[327] | Root tips | 1/20,000 to 1/80,000 | | Chromosome breaks |
| Lindane | *Allium cepa*[141] | Root tips | 0.00125% | | Induced aneuploidy and chromosome fragmentation |
| Lindane | *Allium cepa*[141] | Root tips | 2% to 0.0006% | 0.00125% | Induced C-mitosis |
| Lindane | *Zea mays*[213] | Root tips, stems, coleoptile tissue | Solid particles | | Chromosome aberrations |
| Lindane | *Triticum vulgare*[213] | Root tips, stems, coleoptile tissue | Solid particles | | Chromosome aberrations |
| Lindane | *T. Monococcum*[213] | Root tips, stems, coleoptile tissue | Solid particles | | Chromosome aberrations |
| Lindane | *T. compactum*[213] | Root tips, stems, coleoptile tissue | Solid particles | | Chromosome aberrations |
| Lindane | *Secale cereale*[213] | Root tips, stems, coleoptile tissue | Solid particles | | Chromosome aberrations |

| Pesticide | Organism in Which Tested | Assay System | DOSE Range | Minimum Effective Dose | Biological Effect |
|---|---|---|---|---|---|
| Lindane | *Setaria hatica*[213] | Root tips, stems, coleoptile tissue | Solid particles | | Chromosome aberrations |
| Lindane | *Helianthus annus*[213] | Root tips, stems, coleoptile tissue | Solid particles | | Chromosome aberrations |
| Lindane | *Crepis capillaris*[213] | Root tips, stems, coleoptile tissue | Solid particles | | Chromosome aberrations |
| Lindane | *Vicia faba*[213] | Root tips, stems, coleoptile tissue | Solid particles | | Chromosome aberrations |
| Lindane | *V. sativa*[213] | Root tips, stems, coleoptile tissue | Solid particles | | Chromosome aberrations |
| Lindane | *Brassica nigra*[213] | Root tips, stems, coleoptile tissue | Solid particles | | Chromosome aberrations |
| Linuron | *Rhizobium*[189] | Cells | 0–500 $\mu$g/ml | 200 $\mu$g/ml | Induced forward mutation |
| Linuron | Barley[395] | Anther | 1000 ppm-soaked | | No effect on meiosis ($C_1$) |
| | | | 500 ppm-sprayed | | No effect on meiosis ($C_2$) |
| MH | Barley[197] | Plant cells | $10^{-4}$ M | | Chromosome breaks |
| MH | *Vicia faba*[246] | Root tips | $10^{-4}$ M | | Chromosome breaks, suppresses mitosis |
| MH | *Vicia faba*[87] | Root tips | $10^{-4}$ M | | Chromosome breaks |

| | | | | | |
|---|---|---|---|---|---|
| MH | *Vicia faba*[87] | Root tips | 0.0005 M to 0.00001 M | | Chromosome aberrations |
| MH | *Vicia faba*[120] | Root tips | | | Chromatid breaks |
| MH | *Drosophila melanogaster*[271] | Sperm | 0.4% | | Induced no recessive lethal |
| MH | Guinea pig[27] | Tissue-cultured ear cells | 0.01 M | | No morphological effect |
| MH | Tomato[148] | Root tips | $10^{-2}$ M, $10^{-3}$ M, $10^{-4}$ M | $10^{-4}$ M | Chromosome aberrations |
| MH | Mice[119] | Sperm | 500 mg/kg | | Negative induction of dominant lethals |
| Monuron | Barley[395] | Anther | 1000 ppm-soaked | | Abnormal meiosis ($C_1$) |
| | | | 500 ppm-sprayed | | Abnormal meiosis ($C_2$) |
| Naptalam | Barley[395] | Anther | 1000 ppm-soaked | | Abnormal meiosis ($C_1$) |
| | | | 500 ppm-sprayed | | Abnormal meiosis ($C_2$) |
| Paradichlorobenzene | *Vicia faba*[349] | Root tips | Saturated solution | | Abnormal mitosis<br>Chromosome breaks<br>Chromosome fragmentation |
| Parathion | *Allium cepa*[142] | Root tips | 0.01, 0.005, 0.0075% | | Induced C-mitosis |
| CP | *Allium cepa*[13] | Plant cells | Saturated solution | | Meiotic effect |

| Pesticide | Organism in Which Tested | Assay System | DOSE Range | Minimum Effective Dose | Biological Effect |
|---|---|---|---|---|---|
| Phosphamidon | Barley[395] | Anther | 1000 ppm-soaked | | Slight effect on meiosis ($C_1$) |
| | | | 500 ppm-sprayed | | Slight effect on meiosis ($C_2$) |
| Propham | *Avena sativa*[111] | Root and stem tips | 0.1 to 5.0 ppm | | Mitotic aberrations |
| Propham | *Avena sativa*[238] | Plant cells | 2.5, 5, 10, 20, 40, 80 ppm | | C-mitotic effect |
| Propham | Plant cells[238] | Plant cells | 2.5, 5, 10, 20, 40, 80 ppm | | Nuclear disintegration |
| Propham | *Avena* and *Allium*[111] | Plant cells | | | Anaphase bridges, blocked metaphase, nuclear fragmentation |
| Simazine | Barley[395] | Anther | 1000 ppm-soaked | | Slight effect on meiosis ($C_1$) |
| | | | 500 ppm-soaked | | No effect on meiosis ($C_2$) |
| 2,4,5–T | *Allium cepa*[78] | Root tips | 25 to 500 ppm | 25 ppm | Chromosome aberrations |
| 2,4,5–T | Apricot[51] | Fruit cells | 100 mg/L | | Slight antimitotic effect |

## 3 Bibliography on Mutagenic and Related Effects of Pesticides and Related Compounds

This bibliography was compiled from a manual search of the literature, from the *Environmental Mutagen Society Information Center (EMIC)* registry, and from *MEDLARS*.


1. Adebahr, G.
1966. Changes in the human ovum in E 605 poisoning. *Deutsch. Z. Ges. Gerichtl. Med. 58*: 248–260.

2. Alderson, T.
1957. Culture conditions and mutagenesis in *Drosophila melanogaster*. *Nature 179*: 974–975.

3. ———.
1961. Mechanisms of mutagenesis induced by formaldehyde. The essential role of the 6-amino group of adenylic acid or adenosine in the mediation of the mutagenic activity of formaldehyde. *Nature 191*: 251–253.

4. ———.
1964. The mechanism of formaldehyde-induced mutagenesis. The monohydroxymethylation reaction of formaldehyde with adenylic acid as the necessary and sufficient condition for the mediation of the mutagenic activity of formaldehyde. *Mutation Res. 1*: 77–85.

5. ———.
1960. Significance of ribonucleic acid in the mechanism of formaldehyde-induced mutagenesis. *Nature 185*: 904–907.

6. Alekperov, U. K., Kolomiets, A. F., and Shcherbakov. V. K.
1967. Antimutagenic activity of paraquat. *Dokl. Akad. Nauk SSSR 176*: 199–201.

7. Alexander, M. L.
1967. Genetic damage induced in the sex chromosome and autosomes, with X-ray and ethyleneimine treatments. *Rad. Res. 31*: 614.

8. ———.
1965. Genetic effect of combined X-ray and ethyleneimine treatments. *Genetics 52*: 426.

9–10. ———.
1964. Genetic damage induced in the germ cells of Drosophila spermatogenesis with ethyleneimine. *Genetics 50*: 231–232.

11. ———.
1965. Genetic damage induced by ethyleneimine. *Proc. Natl. Acad. Sci. Wash., 53*: 282–288.

12. Amir, D., and Volcani, R.
1967. The effect of dietary ethylene dibromide (EDB) on the testes of bulls. A preliminary report. *Fertil. Steril. 18*: 144–147.

13. Amer, S. M., and Ali, E. M.
1968. Cytological effects of pesticides. II. Meiotic effects of some phenols. *Cytologia* (Tokyo) *33*: 21–33.

14. Amer. S. M.
1965. Cytological effects of pesticides. I. Mitotic effects of N-methyl-1-naphthyl carbamate, "Sevin". *Cytologia* (Tokyo) *30*: 175–181.

15. Androshchuk, O. F.
1966. Variability of plants evoked by the effect of certain mutagenic chemical substances. *UKR Bot. Zh. 23*: 28–34.

16. Arnason, T. J., Mohammed, L., Koehler, D., and Rennenberg, F. N.
1962. Mutation frequencies in barley after treatment with gamma radiation, ethyleneimine, EMS and maleic hydrazide. *Can. J. Genet. Cytol. 4*: 172–178.

17. ———, and Wakonig, R.
1959. The mutagenic effects of TEM on plants. *Can. J. Genet. Cytol. 1*: 16–20.

18. Atkins, T. D., and Linder, R. L.
1967. Effects of Dieldrin on reproduction of penned hen pheasants. *J. Wildlife Man. 31*: 746–753.

19. Auerbach, C.
1960. Chemical induction of recessive lethals in *N. crassa*. *Microbial Genet. Bull. 17*: 5–6.

20. ———, and Moser, H.
1953. An analysis of the mutagenic action of formaldehyde food. I. Sensitivity of *Drosophila* germ cells. *Z. Vererbungsl. 85*: 547–563.

21. ———.
1953. An analysis of the mutagenic action of formaldehyde food. II. The mutagenic potentialities of the treatment. *Z. Vererbungsl. 85*: 479–504.

22. ———, and Robson, J. M.
1944. Production of mutations by allyl isothiocyanate. *Nature 154*: 81.

23. ———, and Robson, J. M.
1947. Tests of chemical substances for mutagenic action. *Proc. Royal Soc. Edinburgh B. 62*: 284–291.

24. Balazs, I., and Agosin, M.
1968. The effect of 1,1,1-Trichloro-2,2-bis(para-chlorophenyl) ethane on ribonucleic acid metabolism in *Musca Domestica L. Biochim. Biophys. Acta 157*: 1–7.

25. Ball, H. L., Kay, K., and Sinclair, J. W.
1953. Observations on toxicity of Aldrin. I. Growth and estrus in rats. *Arch. Ind. Hyg. Occupational Med. 7*: 292-300.

26. Barnes, J. M.
1964. Toxic hazards and the use of insect chemo-sterilants. *Trans. Royal Soc. Trop. Med. Hyg. 58*: 327–334.

27. ———, Magee, P. N., Boyland, E., Haddow, A., Passey, R. D., Bullough, W. S., Cruickshank, C. N. P.
1957. The non-toxicity of maleic hydrazide for mammalian tissue. *Nature 180*: 62–64.

28. Barratt, R. W., and Tatum, E. L.
1951. An evaluation of some carcinogens as mutagens. *Cancer Res. 11*: 234.

29. ———, and Tatum, E. L.
1958. Carcinogenic mutagens. *Ann. N.Y. Acad. Sci. 71*: 1072–1084.

30. Bartoshevich, Y. E.
1965. The mutagenic effect of nitroso alkyl ureas and ethyleneimine on back-mutation in *Actinomycetes. Dokl. Akad. Nauk SSSR 162*: 193–196.

31. Bartoshevich, Y. E., Filippova, L. M., and Kostyanovskii, R. G.
1966. Investigation of the mutagenic activity of ethyleneimine, urethylanes, nitrosourethylanes and diazoketones derivatives on Actinomyces *strepto-mycini Kras.* and Drosophila melanogaster. *Genetika 4*: 147–155.

32. Basler, E., and Nakazawa, K.
1961. Effects of 2,4-D on nucleic acids of cotton cotyledon tissue. *Bot. Gaz. 122*: 228–232.

33. Bateman, A. J.
1960. The induction of dominant lethal mutations in rats and mice with Triethylenemelamine (TEM). *Genet. Res. (Cambridge) 1*: 381–392.

34. Batikyan, G. G., and Pogosyan, V. S.
1967. Comparative study of the effect of ethylenei-mine and heteroauxin on the mitotic process of the cells of tomato rootlets. *Biol. ZH. Arm. 20(5)*: 3–11.

35. Battacharyja, S. S., and Linskens, H. F.
1955. Concerning the effect of Systox, Meta-Systox, and Pestox on the nucleus and chromosomes of *Vicia faba. Phytopathol. Z. 23*: 233–248.

36. Battaglia, E.

(36. Battaglia, E.)
1949. Cytological activity of ethyl carbamate and cyclohexane carabamic acid. *Caryologia 1*: 229–247; *Biol. Abstract 24*: 3289.

37. Bauch, R. J.
1968. Test results and critical remarks on the application possibility of 5,2′-Dichloro-4′-nitrosalicylanilide ethanolamine in the *Fascioliasis* control. *Arch. Exp. Veterinaermed. 22*: 197–204.

38. Beard, R. L.
1965. Ovarian suppression by DDT and resistance in the house fly *(Musca domestica L.) Entom. Exp. et Appl. 8*: 193–204.

39. Benes, V., and Stram, R.
1968. Mutagenic activity of some pesticides in *Drosophila melanogaster. Industr. Med. & Surgery 37*: 550.

40. Berchev, K., and Minchev, T.
1967. On histologic changes of polysaccharides, desoxyribonucleic acid, ribonucleic acid, and some enzymes in the liver of white rats acted on with zinc phosphide. *Nauch. Tr. Vissh. Med. Inst. Sofiia 46*: 71–76.

41. Bernstein, H., and Miller, A.
1961. Complementation studies with isoleucine-valine mutants of *Neurospora crassa. Genetics 46*: 1039–1052.

42. Beye, F., and Reiff, M.
1962. Über die Kontaktempfindlichkeit sensibler und insektizidresistemeter Drosophila Stamme. I. Mitteilung: Die Wirkung von chlorierten Kohlenwasserstoffen und Phosphorsäureestern. *Anz. Schadlingskunde 35*: 103–109.

43. Bird, M. J.
1952. Chemical production of mutations in *Drosophila*: comparison of techniques. *J. Genet. 50*: 480–485.

44. Bitman, J., Cecil, H. C., and Harris, S. J., and Fries, T. F.
1968. Estrogenic activity of o, p-DDT in the mammalian uterus and ovarian oviduct. *Science 162*: 371–372.

45. Blixt, S.
1964. Studies of induced mutations in peas. VIII. Ethyleneimine and gamma-ray treatments of the variety *Witham wonder. Agri. Hort. Genet. 22*: 171–183.

46. ———, Ehrenberg, L., Gelin, O.
1960. Quantitative studies of induced mutations in peas. III. Mutagenic effect of ethyleneimine. *Agri. Hort. Genet. 18*: 109–123.

47. Bobak M.
1966. The effect of Cycocel (CCC) on the mitotic activity of cells and cell division disorders. *Biologia (Bratislava) 21*: 829–833.

48. ———.
1967. Despiralization of chromosomes and disorders of the chromosome matrix caused by the effect of Propenyl-N-3-fluorophenylcarbamate. *Biologia*

*(Bratislava) 22*: 668–672.

49. Bock, M., and Jackson, H.
1957. The action of Triethylenemelamine on the fertility of male rats. *Brit. J. Pharmacol. 12*: 1–7.

50. Borkovec, A. B.
1962. Sexual sterilization of insects by chemicals. *Science 137*: 1034–1037.

51. Bradley, M. V., and Crane, J. C.
1955. The effect of 2,4,5-Trichlorophenoxyacetic acid on the cell and nuclear size and endopolyploidy in parenchyma of apricot fruits. *Am. J. Bot. 42*: 273–281.

52. Brown, D. M., McNaught, A. D., and Schell, P.
1966. The chemical basis of hydrazine mutagenesis. *Biochem. Biophys. Res. Commun. 24*: 967–971.

53. Bruhin, A., and Wanner, H.
1954. Concerning the action of insecticides on plant mitosis. *Phytopathol. Z. 22*: 327–342.

54. Buiatti, M., and Nuti-Ronchi, V.
1963a. Chromosome Breakage by Triethylenmelamine (TEM) in *Vicia faba* in relation to the mitotic cycle. *Caryologia 16*: 397–403.

55. ———, and Nuti-Ronchi, V.
1963b. Chromosome breakage by Triethylenmelamine (TEM) in *Vicia faba* in relation to the mitotic cycle. *Genetics Today: Proc. XIth Int. Conf. Genet. 1*: 88(A).

56. Burton, D. F.
1950. Anatomy of the cotton leaf and effects induced by 2,4-Dichlorophenoxyacetic acid. *Bot. Gaz. 111*: 325–331.

57. Butenko, R. G., and Baskakov, Y. A.
1961. On mechanism of the effect of maleic hydrazide on plants. *Fiziol. Rastenii 7*: 323–329.

58. Cannon, M. S., and Holcomb, L. C.
1968. The effect of DDT on reproduction in mice. *Ohio J. Sci. 68*: 19–24.

59. Cantwell, G. E., and Henneberry, T. J.
1963. The effects of gamma radiation and Apholate on the reproductive tissues of *Drosophila melanogaster Meigen. J. Insect. Path. 5*: 251–264.

60. Canvin, D. T., and Riesen, G.
1959. Cytological effect of CDAA and IPC on germinating barley and peas. *Weeds 7*: 153–156.

61. Carlson, J. B.
1954. Cytohistological responses of plant meristems to maleic hydrazide. *Iowa State Coll. J. Sci. 29*: 105–128.

62. Carpentier, S., and Fromageot, C.
1950. Activité c-mitotique des isomers $\gamma$ et $\delta$ de l'hexachlorocyclohexane, avec des observations sur l'influence du mesoinositrol et du mesionositophosphate de sodium. *Biochim. et Biophys. Acta 5*: 290–296.

63. Cattanach, B. M.

1957. Induction of translocations in mice by Triethylenemelamine. *Nature 180*: 1364–1365.

64. ———.
1959a. The sensitivity of the mouse testis to the mutagenic action of Triethylenemelamine. *Z. Vererb. 90*: 1–6.

65. ———.
1959b. The effect of Triethylenemelamine on the fertility of female mice. *Int. J. Radiat. Biol. 1*:288–292.

66. ———, and Edwards, R. G.
1957–1958. The effects of TEM on the fertility of male mice. *Proc. Royal Soc. Edinburgh 67*: 54–64.

67. Chamberlain, W. F.
1962. Chemical sterilization of the screw-worm. *J. Econ. Ent. 55*: 240–248.

68. Chang, S. C., and Borkovec, A. B.
1964. Quantitative effects of TEPA, METEPA, and Apholate on sterilization of male house flies. *J. Econ. Ent. 57*: 488–490.

69. Chang, T. H., and Elequin, F. T.
1967. Induction of chromosome aberrations in cultured human cells by ethyleneimine and its relation to cell cycle. *Mutation Res. 4*: 83–89.

70. Chang, T. H., and Klassen, W.
1968. Comparative effects of Tretamine, TEPA, Apholate and their structural analogs on human chromosomes *in vitro. Chromosoma 24*: 314–323.

71. Cheymol, J., Deysson, G., Chabrier, P., and Cheymol, A.
1965. Research on the cytotoxicity of organic combinations of phosphoric acid derivatives. 3. Action of trisubstituted (dialkylaryl and dialkylaralkyl) derivatives of o-phosphoric acid on the human cancer cell cultivated *in vitro. Therapie 20*: 1421–1430.

72. Chrispeels, M. J., and Hanson, J. B.
1962. The increase in ribonucleic acid content of cytoplasmic particles of soybean hypocotyl induced by 2,4-Dichlorophenoxyacetic acid. *Weeds 10*: 123–125.

73. Chury, J., and Slouka, V.
1949. Effects of bromine on mitosis in root-tips of *Allium cepa. Nature 163*: 27–28.

74. Clark, A. M.
1959. Mutagenic activity of the alkaloid Heliotrine in *Drosophila. Nature 183*: 731–732.

75. Clowes, G. H. A.
1951. The inhibition of cell division by substituted phenols with special reference to the metabolism of dividing cells, *Ann. N.Y. Acad. Sci. 51*: 1409–1431.

76. Cohn, N. S.
1961. Production of chromatid aberrations of diepoxybutane and an iron chelator. *Nature 192*: 1093–1094.

77. Cressman, A. W.
1963. Response of citrus red mite to chemical sterilants. *J. Econ. Ent. 56*: 111–112.

78. Croker, B. H.
1953. Effects of 2,4-Dichlorophenoxyacetic acid and 2,4,5-Trichlorophenoxyacetic acid on mitosis in *Allium cepa*. *Bot. Gaz. 114*: 274–283.

79. Crystal, M. M.
1963. The induction of sexual sterility in the screw-worm fly by antimetabolites and alkylating agents. *J. Econ. Ent. 56*: 468–473.

80. ———.
1964a. Sexual sterilization of screw-worm flies by the biological alkylating agents, Tretamine and Thiotepa. *Exptl. Parasitol. 15*: 249–259.

81. ———.
1964b. Chemosterilant efficiency of bis(1-aziridinyl)-phosphinyl carbamates in screw-worm flies. *J. Econ. Ent. 57*: 726–731.

82. ———, and LaChance, L. E.
1963. The modification of reproduction in insects treated with alkylating agents. I. Inhibition of ovarian growth and egg production and hatchability. *Biol. Bull. 125*: 270–279.

83. D'Amato, F.
1948. Sull'attivita colchicino-mitotica e su altri effetti citologici del 2,4-diclorofenossiacetato di sodio. *Atti Accad. Nazl. Lincci, Rend., Classe sci. fis., mat. e nat. (8)4*: 570–578.

84. ———.
1950. The quantitative study of mitotic poisons by the *Allium cepa* test: Data and problems. *Protoplasma 39*: 423–433.

85. Danes, B., and Leinfelder, P.
1951. Cytological and respiratory effects of cyanide on tissue cultures. *J. Cell Comp. Phys. 37*: 427–446.

86. Darlington, C. D.
1947. The chemical breaking of chromosomes. *Heredity 1*: 187–221.

87. ———, and McLeish, J.
1951. Action of maleic hydrazide on the cell. *Nature 167*: 407–409.

88. Daugherty, J. W., Lacey, D. E., and Korty, P.
1962. Some biochemical effects of Lindane and Dieldrin on vertebrates. *Aerospace Med. 33*: 1171–1176.

89. Davis, D. E.
1962. Gross effects of Triethylenemelamine on gonads of starlings. *Anat. Record 142(3)*: 353–358.

90. Davis, K. J., and Fitzhugh, O. G.
1962. Tumorigenic potential of Aldrin and Dieldrin for mice. *Toxicol. Appl. Pharmacol. 4*: 187–189.

91. Dean, A. C. R., and Law, H. S.
1964. The action of 2,4-Dichlorophenoxyacetic acid in *Aerobacter aerogenes*. *Ann. Botany (London) 28*: 703–710.

92. Deichmann, W. B., and Keplinger, M. L.
1966. Effect of combinations of pesticides on reproduction of mice. *Toxicol. Appl. Pharmacol. 8*: 337–338.

93. ———, Keplinger, M. L., and Glass, E. M. Synergism among oral carcinogens. IV. Results of the simultaneous feeding of four tumorigens to rats. *Abst. 6th Ann. Toxicol. Soc. Atlanta, Georgia.* Mar. 23–25.

94. Deviovanni-Donnelly, R., Kolbye, S. M., and Greeves, P. D.
1968. The effects of IPC, CIPC, Sevin and Zectran on *Bacillus subtilis, Experientia 24*: 80–81.

95. ———, and DiPaolo, J.
1967. The effect of carbamates on *Bacillus subtilis. Mutation Res. 4*: 543–551.

96. DeWitt, J. B.
1956. Chronic toxicity to quail and pheasants of some chlorinated insecticides. *J. Agr. Food Chem. 4*: 863–866.

97. ———.
1955. Effects of chlorinated hydrocarbon insecticide upon quail and pheasants. *J. Agr. Food Chem. 3*: 672–676.

98. Dickey, F. H., Cleland, G. H., and Lotz, C.
1949. Organic peroxides in the induction of mutations *Proc. Natl. Acad. Sci. U.S. 35*: 581–586.

99. Doxey, D. C.
1949. The effect of Isopropylphenyl carbamate on mitosis in rye and onion. *Ann. Bot. 13*: 329–335.

100. ———, and Rhodes, A.
1949. The effect of plant growth regulator 4-Chloro-2-methylphenoxyacetic acid on mitosis in the onion. *(Allium cepa). Ann. Botany (N.S.) 13*: 105–111.

101. ———.
1951. The effects of the gamma isomer of benzene hexachloride (Hexachlorocyclohexane) on plant growth and on mitosis. *Ann. Botany (N.S.) 15*: 47–52.

102. Dubinin, N. P., Mitrofanov, Y. A., and Manuilova, E. S.
1967. Analysis of the Thiotepa mutagenic effect on human tissue culture cells. *Izv. Akad. Nauk. SSSR (Biol.) (4)*, 477.

103. Dubrowin, K. P.
1959. Cytohistological response of wheat and wild oats to Carbyne (4-chloro-2-butynyl-N-[3-chlorophenyl]carbamate). *Proc. NCWCC 16*: 15.

104. Duggan, R. E., and Weatherwax, J.
1967. Dietary intake of pesticide chemicals. *Science 157*: 1006–1010.

105. ———.
1966. Effect of some insecticides on the hatching of hen's eggs. *Nature 212*: 1062–1063.

106. Dumachie, J. F., and Fletcher, W. W.
1967. Effect of some herbicides on the hatching rate of hens' eggs. *Nature 215*: 1406–1407.

107. Eden, W. G.
1951. Toxicity of Dieldrin to chickens. *J. Econ. Ent. 44*: 1013.

108. Ehrenberg, L., Lundqvist, U., and Giunner, S.

1958. The mutagenic action of ethylene imine in barley. *Hereditas 44*: 330–336.

109. ———, and Gustafsson, Å.
1957. On the mutagenic action of Ethylene oxide and diepoxybutane in barley. *Hereditas 43*: 595–602.

110. ———, and Lundqvist, U.
1959. The mutagenic effects of ionizing radiation and reactive ethylene derivatives in barley. *Hereditas 45*: 351–368.

111. Ennis, W. B., Jr.
1948. Some cytological effects of o-Isopropyl-N-phenylcarbamate upon *Avena*. *Am. J. Bot. 35*: 15–21.

112. Epstein, J., Rosenthal, R. W., and Ess, R. J.
1955. Use of 4-(4-nitrobenzyl)pyridine as analytical reagent for ethyleneimes and alkylating agents. *Anal. Chem. 27*: 1435–1439.

113. Epstein, S. S., Andrea, J., Clapp, P., and Mackintosh, D.
1967. Enhancement by Piperonyl butoxide of acute toxicity due to Freons, benzo[a]pyrene, and Griseofulvin in infant mice. *Toxicol. Appl. Pharmacol. 11*: 442–448.

114. ———, Arnold, E., Steinberg, K., Mackintosh, D., Shafner, H., and Bishop, Y.
1970. Mutagenic and antifertility effects of TEPA and METEPA in mice. *Toxicol. Appl. Pharmacol. 17*: 23–40.

115. ———, Bass, W., Arnold, E., and Bishop, Y.
1970. Mutagenicity of trimethylphosphate in mice. *Science 168*: 584–586.

116. ———, Bass, W., Arnold, E., and Bishop, Y.
1970. The failure of caffeine to induce mutagenic effects or to synergize the effects of known mutagens in mice. Food Cos. *Toxicol. 8*: 381–401.

117. ———, Joshi, S., Andrea, J., Clapp, P., Falk, H., and Mantel, N.
1967. The synergistic toxicity and carcinogenicity of Freons and Piperonyl butoxide. *Nature 214*: 526–528.

118. ———, and Mantel, N.
1968. Hepatocarcinogenicity of maleic hydrazide. *Int. J. Cancer 3*: 325–335.

119. ———, and Shafner, H.
1968. Chemical mutagens in the human environment. *Nature 219*: 385–387.

120. Evans, H. J., and Scott, D.
1964. Influence of DNA synthesis on the production of chromatid aberrations by X-rays and maleic hydrazide in *Vicia faba*. *Genetics 49*: 17–38.

121. Faberge, A. C.
1955. Types of chromosome aberrations induced by Ethylene oxide in maize. *Genetics* Princeton *40*: 571.

122. Fabro, S., Smith, R. L., and Williams, R. J.
1966. Embryotoxic activity of some pesticides and drugs related to phthalimide. *Food Cos. Toxicol. 3*: 587–590.

123. Fahmy, F. Y.

1951. Cytogenetic analysis of the action of some fungicide mercurials. Ph.D. thesis, Inst. Genet. Univ. of Lund, Sweden.

124. Fahmy, O. G., and Bird, M. J.
1952. Chromosome breaks among recessive lethals induced by chemical mutagens in Drosophila. *Heredity 6* (Suppl.): 149–159.

125. ———, and Fahmy, M. J.
1954. Cytogenetic analysis of the action of carcinogens and tumour inhibitors in *Drosophila melanogaster*. II. The mechanism of induction of dominant lethals by 2:4:6-tri(ethyleneimino)-1:3:5-triazine. *J. Genet. 52*: 603–619.

126. ———.
1964. The chemistry and genetics of the alkylating chemosterilants. *Trans. Royal Soc. Trop. Med. Hyg. 58*: 318–326.

127. ———.
1955. Cytogenetic analysis of the action of carcinogens and tumour inhibitors in *Drosophila melanogaster*. IV. The cell stage during spermatogenesis and the induction of intra- and inter-genic mutations by 2:4:6-tris(ethylenimino) – 1:3:5-triazine. *J. Genet. 53*: 563–584.

128. Fan, D. and Maclachlan, G. A.
1967. Massive synthesis of ribonucleic acid and cellulose in the pea epicotyl in response to indoleacetic acid, with and without concurrent cell divisions. *Plant Physiol. 42*: 1114–1122.

129. Ferguson, V.
1965. Some aspects of the storage effect on chromosome breakage produced by Triethylene melamine in *Drosophila melanogaster*. B. Sc. Thesis, University of Edinburgh.

130. Fletcher, K.
1967. Production and viability of eggs from hens treated with Paraquat. *Nature 215*: 1407–1408.

131. Friesen, H. A., Baenziger, H., and Keys, C. H.
1964. Morphological and cytological effects of Dicamba on wheat and barley. *Can J. Plant Sci. 44*: 288–294.

132. Frizzi, G.
1967. Biological effects of Apholate on insects. *Wiad. Parazyt. 13*: 373–377.

133. Fucik, V., Michaelis, A., and Rieger, R.
1963. Different effects of BUDR on the induction of chromatid aberrations by means of TEM and maleic hydrazide. *Biochem. Biophys. Res. Commun. 13*: 366–371.

134. Fujii, K. and Epstein, S.S.
1969. Carcinogenicity of food additives, pesticides, and drugs after parenteral administration in infant mice. Abst. 8th Annual Meeting Society of Toxicology. Williamsburg, Virginia. Mar. 10–12, 1969.

135. Gabliks, J., Bantung-Jurilla, M., and Friedman, L.

1967. Responses of cell cultures to insecticides. IV. Relative toxicity of several organophosphates in mouse cell cultures. *Proc. Soc. Exp. Biol. Med. 125*: 1002–1005.

136. Gaines, T. B.
1960. The acute toxicity of pesticides to rats. *Toxicol. Appl. Pharmacol. 2*: 88–99.

137. Galley R. A. E.
1952. Problems arising from the use of chemicals in food. The toxicity of residual agricultural chemicals. *Chem. Ind.*, No. 16 342–344.

138. Ghadiri, M., Greenwood, D. A., and Binns, W.
1967. Feeding of Malathion and Carbaryl to laying hens and roosters. *Tox. Appl. Pharm. 10*: 392.

139. Ghatnekar, M. V.
1964. Primary effects of different mutagens and the disturbances induced in the meiosis of $X_1$ and $X_2$ of *Vicia faba. Caryologia 17*: 219–244.

140. Gifford, E. M., Jr.
1956. Some anatomical and cytological responses of barley to maleic hydrazide. *Am. J. Bot. 43*: 72–80.

141. Gimenez-Martin, G. and Lopez-Saez, J.
1961. Autogenesis por accion del gamma-hexaclorociclohexano sobre indivision celular. *Phyton 16*: 45–55.

142. ———.
1962. Accion a-Mototica del Parathion. *Phyton 18*: 23–26.

143. Giovanni-Donnelly, R. D. E., Kolbye, S. M., and Dipaolo, J. A.
1967. The effect of carbamates on *Bacillus subtilis. Mutation Res. 4*: 543–551.

144. Good, E. E., and Ware, G. W.
1969. Effects of insecticides on reproduction in the laboratory mouse. *Toxicol. Appl. Pharmacol. 14*: 201–203

145. Good, E. E., Ware, G. W., and Miller, D. F.
1965. Effects of insecticides on reproduction in the laboratory mouse. I. Kepone. *J. Econ. Entol. 58*: 754–757.

146. Gowdey, C. W., Graham, R. A., Sequin, J. J., and Stauraky, G. W.
1954. The pharmacological properties of the insecticide Dieldrin. *Can. J. Biochem. Physiol. 32*: 498–503.

147. Graf, G. E.
1957. Chromosome breakage induced by X-rays, maleic hydrazide and its derivatives. *J. Heredity 48*: 155–159.

148. Grant, W. F. and Harney, P. M.
1960. Cytogenetic effects of maleic hydrazide treatment of tomato seed. *Can. J. Genet. Cytol. 2*: 162–174.

149. Greulach, V. A., and Atchison, E.
1956. Inhibition of growth and cell division in onion roots by maleic hydrazide. *J. Bot. 35*: 15–21.

150. ———, and Haesloop, J. G.
1953. Some cytological and anatomical effects of maleic hydrazide. *J. Elisha Mitchell Sci. Soc. 69*: 88–89.

151. ———.
1954. Some effects of maleic hydrazide on internode elongation, cell enlargement, and stem anatomy. *Am. J. Bot. 41*: 44–50.

152. Grigorov, I., Petrunova, S., Dzherova, A., and Slokosk, A. L.
1967. Study of processes of respiration and amino acids synthesis in variants of *Aspergillus tamarii* obtained by the action of ethyleneimine. *Izv. Mikrobiol. Inst. 19*: 115–121.

153. Grosch, D. S.
1967. Poisoning with DDT: Effect on reproductive performance of *Artemia. Science 155*: 592–593.

154. ———, and Valcovic, L. R.
1964. Genetic analysis of the effects of Apholate, *Bull. Ent. Soc. Amer. 10*: 163.

155. ———.
1967. Chlorinated hydrocarbon insecticides are not mutagenic in *Bracon hebetor* tests. *J. Econ. Ent. 60*: 1177–1179.

156. Gurenwedel, D. W., and Davidson, N.
1966. Complexing and denaturation of DNA by methylmercuric hydroxide. I. Spectrophotometric studies. *J. Mol. Biol. 21*: 129–144.

157. Haber, A. H.
1962. Effects of indoleacetic acid on growth without mitosis and on mitotic activity in the absence of growth by expansion. *Plant Physiol. 37*: 18–26.

158. Hampel, K. E., and Gerhatz, R. H.
1965. Strukturanomalien der chromosomen menschlicher Leukozyten *in vitro* durch Triethylenemelamin. *Exp. Cell. Res. 37*: 251–258.

159. Hashimoto, Y., Makita, T., Miyata, H. Noguchi, T., and Ohta, G.
1968. Acute and subchronic toxicity of a new Fluorine pesticide. *N*-methyl-*N*-(1-naphthyl)-fluoroacetamide. *Toxicol. Appl. Pharmacol. 12*: 536–547.

160. Havertz, D. S., and Curtin, T. J.
1967. Reproductive behavior of *Aedes aegypti* (L.) sublethally exposed to DDT. *J. Med. Entom. 4*: 143–145.

161. Hayes, W. J.
1959. The toxicity of Dieldrin to man. Report of a survey. *Bull. World Health Organ. 20*: 891–912.

162. Hellenbrand, K.
1966. Die Wirkung kombinierter Anwendung insektizider Phosphorsaueester und carbamate auf *Drosophila melanogaster*. *Ent. Exp. & Appl. 9*: 232–246.

163. Herrick, R. B., and Sherman, M.
1964. Effect of an alkylating agent, Apholate, on the

chicken. *Poultry Sci. 43*: 1327–1328.

164. Herskowitz, I. H.
1951. Mutation rate in *Drosophila melanogaster* males treated with formaldehyde and 2,4-dinitrophenol. *Genetics 36*: 554–555.

165. ———.
1956. The incidence of chromosomal rearrangements and recessive lethal mutations following treatment of mature *Drosophila* sperm with 2:4:6-tri (ethyleneimino) 1:3:4-triazine. *Genetics 40*: 574.

166. ———.
1956. Mutagenesis in mature *Drosophila* spermatozoa by "Triazine" applied in vaginal douches. *Genetics 41*: 605–609.

167. Higashi, K., Matsuhisa, T., Kitao, T., and Sakamotu, Y.
1968. Selective suppression of nucleolar RNA metabolism in the absence of protein synthesis. *Biochim. Biophys. Acta 166*: 388–393.

168. Hiltibran, R. C.
1967. Effects of some herbicides on fertilized fish eggs and fry. *Trans. Amer. Fish. Soc. 96*: 414–416.

169. Hindmarsh, M. M.
1952. A critical consideration of c-mitosis with reference to the effects of nitrophenols. *Proc. Linnean Soc. Wales 76*: 158–163.

170. Hodge, H. E., Boyce, A. M., Deichmann, E. B., and Kraybill, H. F.
1967. Toxicology and no-effect levels of Aldrin and Dieldrin. *Toxicol. Appl. Pharmacol. 19*: 613–675.

171. ———, Downs, W. L., and Smith, D. W.
1968. Oral toxicity of Linuron (3-[3,4-dichlorophenyl]-1-methoxy-1-methylurea) in rats and dogs. *Food Cos. Toxicol. 6*: 171–183.

172. Hoffman, W. S., Fishbein, W. I., and Andelman, M. B.
1964. The pesticide content of human fat tissue. *Arch. Environ. Health 9*: 387–394.

173. Horvath, L.
1966. Embryopathy caused by insecticide. *Orv. Hetil. 107*: 2001–2004.

174. Hunter, C. G., Rosen, A., Williams, R. T., Reynolds, J. G., and Worden, A. N.
1960. Studies on the fate of aldrin, dieldrin, and endrin in the mammal. *Mededel. Landbouwhogeschool Opzoekingrstat. Staat. Genet. 25*: 1296–1307.

175. Hunter, P. E., Cutkomp, L. J., and Kolkaila, A. M.
1958. Reproduction in DDT- and Diazinon-treated houseflies. *J. Econ. Ent. 51*: 579–582.

176. Innes, J. R. M., Ulland, B. M., Valerio, M. G., Petrucelli, L., Fishbein, L., Hart, E. R., Pallotta, A. J., Bates, R. D., Falk, H. L., Galt, J. J., Klein, M., Mitchell, I., and Peters, J.
1969. Bioassay of pesticides and industrial chemicals for tumorigenicity in mice: A preliminary note. *J. Nat. Cancer Inst. 42*: 1101–1116.

177. Isenberg, F. M. R., Odland, M. L., Popp, H. W., and Jensen, C. O.
1951. The effect of maleic hydrazide on certain dehydrogenases in tissues of onion plants. *Science 113*: 58–60.

178. Iyer, C. P. A., and Randhawa, G. S.
1966. Induction of pollen sterility in grapes *(Vitts vinifera)*. *Vitis Ber. Rebenforsch. 5*: 433–445.

179. Iyer, V. N., and Szybalski, W.
1958. The mechanism of chemical mutagenesis. I. Kinetic studies on the action of Triethylene melamine (TEM) and azaserine. *Proc. Nat. Acad. Sci. U.S. 44*: 446–456.

180. ———.
1959. Mutagenic effect of azaserine in relation to azaserine resistance in *Escherichia coli*. *Science 129*: 839–840.

181. Janovski, I., Dumanovic, J., Denic, M., and Perakliev, D.
1967. Factors influencing the sensitivity of wheat and cotton seeds to ethyleneimine (EI) and ethyl methanesulfonate (EMS). *Cont. Agric. 15*: 13–23.

182. Jefferies, D. J.
1967. The delay in ovulation produced by pp[1]-DDT and its possible significance in the field. *IBIS 109*: 266–272.

183. Jensen, K. A., Kirk, I., Kolmark, G., and Westergaard, M.
1951. Chemically induced mutations in *Neurospora*. *Cold Spring Harbor Symp. Quant. Biol. 16*: 245–261.

184. Judson, Charles L.
1967. Alteration of feeding behavior and fertility in *Aedes aegypti* by the chemosterilant Apholate. *Ent. Exp. Appl. 10*: 387–394.

185. Juhasz, J., Balo, J., and Szende, B.
1966. Tumor inducing effect of hydrazine in mice. *Nature 210*: 1377.

186. Kaplan, W. D., and Seecof, R.
1966. The mutagenic action of Aramite, an acaricide *Drosophila Information Service No. 41*: 101.

187. Kaszubiak, H.
1966. The effect of herbicides on *Rhizobium*. I. Susceptibility of *Rhizobium* to herbicides. *Acta Microbiol. Polon. 15*: 357–364.

188. ———.
1968. The effect of herbicides on *Rhizobium*. II. Adaptation of *Rhizobium* to Afalon, Aretit and Liro-betarex. *Acta Microbiol. Polon. 17*: 41–50.

189. ———.
1968. The effect of herbicides on *Rhizobium*. III. Influence of herbicides on mutation. *Acta. Microbiol. Polon. 17*: 51–58.

190. Kaufman, P. B.
1953. Gross morphological responses of the rice plant to 2,4-D. *Weeds 2*: 223–253.

191. Kawai, T., and Sato, H.

1965. Studies on artificial induction of mutations in rice by chemicals. I. Chlorophyll mutations by ethyleneimine and Ethylene oxide treatment. *Bull. Natl. Inst. Agr. Japan D. 13*: 133–162.

192. Keplinger, M. L., Deichmann, W. B., and Sala, F. 1968. Effects of combinations of pesticides on reproduction in mice. *Intern. J. of Ind. Med. & Surgery 37*: 525.

193. Khera, K. S., and Clegg, D. J. 1969. Perinatal toxicity of pesticides. *Can. Med. Assn J. 100*: 167–172.

194. Khishin, A. F. 1956. Induction of mutations by formaldehyde solutions in *Drosophila melanogaster*. *Am. Nat. 90*: 377–380.

195. ———. 1964. The requirement of adenylic acid for formaldehyde mutagenesis. *Mutation Res. 1*: 202–205.

196. Khvostova, V. V., Mozhaeva, V. S., Agaes, N. S., and Veleva, S. A. 1965. Mutants induced by ionizing radiations and ethyleneimine in winter wheat. *Mutation Res. 2*: 339–344.

197. Kihlman, B. A. 1959. On the radiomimetic effects of cupferron and potassium cyanide. *J. Biophys. Biochem. Cytol. 5*: 351–353.

198. Kilbey, B. J., and Kølmark, H. G. 1968. A mutagenic after-effect associated with ethylene oxide in *Neurospora Crassa*. *Molec. Gen. Genet. 101*: 185–188.

199. Kilgore, W. W., and Painter, R. R. 1964. Effect of the chemosterilant Apholate on the synthesis of cellular components in developing housefly eggs. *Biochem. J., 92*: 353–357.

200. Kimball, R. F. 1965. The induction of reparable premutational damage in *Paramecium aurelia* by the alkylating agent triethylene melamine. *Mutation Res. 2*: 413–425.

201. Kitselman, C. H. 1953. Long term studies on dogs fed Aldrin and Dieldrin in sublethal dosages, with reference to the histopathological findings and reproduction. *J. Am. Vet. Med. Assoc. 123*: 28–30.

202. Klassen, Waldemar, and Chang, T. H. 1966. Tris- (1-aziridinyl) phosphine oxide: Caution on use. *Science 154*: 920.

203. Knaak, J. B., Sullivan, L. J., and Wills, J. H. 1967. Metabolism of carbaryl in man. *Abst. 6th Ann. Mtg. Soc. Tox.* Atlanta.

204. Kølmark, G. F. 1956. Mutagenic properties of certain esters of inorganic acids investigated by the *Neurospora* back-mutation test. *Comptes Rend. Trav. lab. Carlsberg ser. Physiol. 26*: 205–220.

205. ———, and Westergaard, M.

(205. ———, and Westergaard, M.)
1953. Further studies on chemically induced reversions at the adenine locus of *Neurospora*. *Hereditas 39*: 209–224.

206. ———, and Auerbach, C.
1960. Mutagenic after-effect in *Neurospora* treated with diepoxybutane. *Microbiol. Gen. Bull. 17*: 24–25.

207. Kølmark, H. G., and Kilbey, B. J.
1962. An investigation into the mutagenic after-effect of butadiene diepoxide using *Neurospora crassa*. *Z. Vererb. 93*: 356–365.

208. ———, and Kondo, S.
1963. Kinetic studies of chemically induced reverse mutations in *Neurospora crassa*. *Proc. Intern. Congr. Genet. 11th, The Hague 1*: 61–62.

209. Kostoff, D.
1938. Irregular mitosis and meiosis induced by acenaphthene. *Nature 141*: 1144–1145.

210. ———.
1939. Effect of the fungicide "Granosan" on atypical growth and chromosome doubling in plants. *Nature 144*: 334.

211. ———.
1940. Atypical growth, abnormal mitosis and polyploidy induced by Ethyl-mercury-chloride. *Phytopathol. Z. 23:* 90–96.

212. ———.
1948. Cytogenetic changes and atypical growth induced by hexachlorocyclohexane ($C_6H_6CL_6$). *Current Science 17*: 294–295.

213. ———.
1949. Induction of cytogenetic changes and atypical growth by hexachlorocyclohexane. *Science 109*: 467–468.

214. Kraybill, H. F., ed.
1969. Biological effects of pesticides in mammalian systems. *Ann. N.Y. Acad. Sci. 160*: 1–42.

215. Kruglyakova, K. E., Ulanov, E. P., and Emmanuel, N. M.
1966. Kinetic characteristics of the action of chemical mutagens (ethyleneimine derivatives) on deoxyribonucleic acid. *Dokl. Akad. Nauk. SSSR 161/3*: 718–720.

216. LaBreque, G. C.
1961. Studies with three alkylating agents as house fly sterilants. *J. Econ. Ent. 54*: 684–689.

217. LaChance, L. E., and Crystal, M. M.
1963. The modification of reproduction in insects treated with alkylating agents. II. Differential sensitivity of oocyte meiotic stages to the induction of dominant lethals. *Biol. Bull 125*: 280–288.

218. LaChance, L. E., and Leverich, A. P.
1968. Chemosterilant studies on *Bracon* (Hymenoptera: Braconidae) sperm. I. Sperm inactivation and dominant lethal mutations. *Ann. Ent. Soc. Amer. 61(1)*: 164–173.

219. LaChance, L. E., and Riemann, J. G.

1964. Cytogenetic investigations on radiation and chemically induced dominant lethal mutations in oocytes and sperm of the screw-worm fly. *Mutation Res. 1*: 318–333.

220. Landa, V., and Rezabova, B.
1965. The effect of chemosterilants on the development of reproductive organs in insects. *Proc. XII Int. Congr. Entom.* London: 516–517.

221. Landa, Z.
1959. $\gamma$-Hexachlorocyclohexane as a polyploidy-inducing agent. *Biol. Plant Acad. Sci. Bohemoslov. 1*: 151–156.

222. Legator, M. S., Kelly, F. J., Green, S., and Oswald, E. J.
1969. Mutagenic Effects of Captan. *Ann. N.Y. Acad. Sci. 160*: 344–351.

223. Levan, A.
1940. The effect of acenaphthene and colchicine on mitosis of *Allium* and *Colchicum. Hereditas 26*: 262–276.

224. ———.
1948. The influence on chromosomes and mitosis of chemicals, as studied by the *Allium* test. *Proc. 8th Internatl Cong. Genet.* Stockholm 325–337.

225. ———, and Tjio, J. H.
1948. Chromosome fragmentation induced by phenols. *Hereditas 34*: 250–251

226. Lilly, L. J., and Thoday, M.
1956. Effects of cyanide on the roots of *Vicia faba. Nature 177*: 338–339.

227. Lingens, F.
1961. Mutagene Wirkung von Hydrazine auf *Escherichia coli*-Zellen. *Naturwissenschaften 48*: 480.

228. Lorkiewicz, L., and Szybalski, W.
1961. Mechanism of chemical mutagensis. IV. Reaction between triethylene melamine and nucleic acid components. *J. Bact. 82*: 195.

229. Loveless, A.
1952. Chemical and biochemical problems arising from the study of chromosome breakage by alkylating agents. *Heredity 6* (Suppl.): 293–298.

230. Lüers, H.
1953. Untersuchungen über die Mutagenität des Triethylenemelamin (TEM) an *Drosophila melanogaster. Arch. Geschwulstforsch. 6*: 77–83.

231. ———.
1959. The mutagenicity of Triethylene thiophosphoramide (Thio-TEPA). *Dros. Inf. Ser. 33*: 145.

232. ———, and Rohrborn, G.
1963. The mutagenic activity of ethyleneimine derivatives with different numbers of reactive groups. *Proc. of the XIth Internatl. Congress of Genetics,* Sept. 1, 1963.

233. Lwoff, A., Jacob, F., Ritz, E., and Gagi, M.
1952. Induction de la production de bacteriophages et d'une colicine par des peroxydes les ethyleneimines

et les halogenoalcoylamines. *Compt. Rend, Hebd. Séanc. Acad. Sci.* (Paris) *234*: 2308–2310.

234. MacFarlane, E. W. E.
1950. Somatic mutations produced by organic mercurials in flowering plants. *Genetics 35*: 122–123.

235. Mackie, A., and Parnell, I. W.
1967. Some observations on *Taeniid* ovicides—the effects of some organic compounds and pesticides on activity and hatching. *J. Helminth. 41*: 167–210.

236. Malling, H. V.
1969. Ethylene dibromide: a potent pesticide with high mutagenic activity. *Genetics 61*: (Suppl.): 39.

237. Mann, J. D., Jordan, L. S., and Day, B. S.
1965. The effects of carbamate herbicides on polymer synthesis. *Weeds 13*: 63–66.

238. ———, and Storey, W. B.
1966. Rapid action of carbamate herbicides upon plant cell nuclei. *Cytologia 31*: 203–207.

239. Markarian, D. S.
1966. Cytogenetic effect of some chlorine-containing organic insecticides on mouse bone-marrow cell nuclei. *Genetika 1*: 132–137.

240. ———.
1967. Effect of Dieldrin on the mitosis in *Crepis capillaris* sprouts. *Genetika 3*: 55–58.

241. Matthies, F.
1966. The triad of agnotia, facial paralysis, and cardiac anomaly not due to Thalidomide. *J.A.M.A. 195*: 695–696.

242. Mauldin, J. K., Hammer, A. L., and Brezzel, J. R.
1966. The effect of insecticides on egg production in the boll weevil (*A. grandis*). *J. Ga. Ent. Soc. 1*: 15–19.

243. McCarthy, R. E., and Epstein, S. S.
1968. Cytochemical and cytogenetic effects of maleic hydrazide on cultured mammalian cells. *Life Sciences 7*: Part 2, 1–6.

244. McGahen, J. W., and Hoffmann, C. E.
1966. Absence of mutagenic effects of 3- and 6-alkyl-5-bromouracil herbicides on a bacteriophage. *Nature 209*: 1241–1242.

245. McLean, R. G.
1967. Chemical control of reproduction in confined populations of pigeons (*Columba livia*). *Diss. Abstr. Sect. B. 28(3)*: 1277B.

246. McLeish, J.
1952. The action of maleic hydrazide in *Vicia*. *J. Heredity 6* (Suppl.): 125–147.

247. McMahon, R. M.
1956. Mitosis in polyploid somatic cells of *Lycopersicon esculentum*. *Caryologia 8*: 250–256.

248. ———, Witkus, E. R., and Berger, C. A.
1960. Some cytological characteristics of 2, 4-D induced mitosis in polyploid cells of *Solanum tuberosum*. *Caryologia 12*: 398–403.
*Caryologia 12*: 398–403

249. Mengle, D., Hale, W., and Rappolt, R. T.
1966. Hematologic abnormalities and pesticides. *Calif. Med. 107*: 251–253.

250. Michalek, S. M., and Brockman, H. E.
1969. A test of mutagenicity of Shell "no-pest strip insecticide" in *Neurospora crassa. Neurospora Newsletter 14*: 8.

251. Miller, R. M.
1967. Prenatal origins of mental retardations: Epidemiological approach. *J. Pediat. 71(3)*: 455–458.

252. Mitchell, J. S., and Simon-Reuss, I.
1952. Experiments on the mechanism of action of tetrasodium 2-methyl-1:4-naphthohydroquinone diphosphate as a mitotic inhibitor and radiosensitizer, using the method of tissue culture. Experimental methods and quantitative results. *Brit. J. Cancer 6*: 305–316,

253. Mitlin, N., Butt, B. A., and Shortino, T. J.
1957. Effect of mitotic poisons on house fly oviposition. *Physiol. Zool. 30*: 133–136.

254. Mohamed, A. H., Smith, J. D., and Applegate, H. G.
1966. Cytological effects of hydrogen fluoride on tomato chromosomes. *Can. J. Gen. Cytol. 8*: 575–583.

255. Morgan, P. B.
1967. Effect of Hempa on the ovarian development of house flies (*Musca domestica* L.) *Ann. Ent. Soc. Amer. 60*: 812–818.

256. ———, and LaBreque, G. C.
The effect of apholate on the ovarian development of house flies. *J. Econ. Ent. 55*: 626–628.

257. ———.
1964. Effect of TEPA and METEPA on ovarian development of house flies. *J. Econ. Ent. 57*: 896–899

258. Morris, R. D.
1968. Effects of Endrin feeding on survival and reproduction in the deer mouse, *Peromyscus mainculatus. Can. J. Zool. 46*: 951–958.

259. Morrison, J. W.
1962. Cytological effects of the herbicide "Avadex." *Can. J. Plant Sci. 42*: 78–81.

260. Moutschen-Dahmen, J.
1962. Effets combines du myleran et de l'hydrazide maleique sur les chromosomes de *Vicia faba* et leur importance dans l'analyse du patrimoine héréditaire. *Lejeunia* N. S. Nr. 10: 1–9.

261. ———, Dahmen, M., and Gillel, C.
1956. Sur les modifications induites par les hydrazides soumises à des agents chimiques mutagènes. *Rev. Cytol. Biol. Veg. 24*: 265–275.

262. ———.
1956. Sur les modifications induites par les hydrazides maléique et isonicotinique dans les antheridées de *Chara vulgaris* L. *La Cellule 58*: 63–70.

263. Moutschen-Dahmen, J., Moutschen-Dahmen, M., and Ehrenberg, L.
1968. Note on the chromosome breaking activity of ethylene oxide and ethyleneimine. *Hereditas-Genet. Ark. 60*: 267–269.

264. Muehling, G. N., Vant Hof, J., Wilson, G. B., and Brigsby, B. H.
1960. Cytological effects of herbicidal substituted phenols. *Weeds 8*: 173–181.

265. Muirhead-Thomson, R. C. and Merryweather, J.
1969. Effect of larvicides on *Simulium* eggs. *Nature 221*: 858–859.

266. Naber, E. C., and Ware, G. W.
1965. Effect of Kepone and Mirex on reproductive performance in the laying hen. *Poultry Sci. 44*: 875–880.

267. Nafei, H., and Auerbach, C.
1964. Mutagenesis by formaldehyde food in relation to DNA replication in *Drosophila* spermatocytes. *Z. Vererbungsl. 95*: 351–367.

268. Nagasawa, S., and Shinohara, H.
1964. Sterilizing effect of Metepa on the azuki bean weevil, *Callosobruchus chinensis L.*, with special reference to the hatching of the eggs deposited by treated weevils. *Japan. J. Appl. Ent. Zool. 8*: 123–128.

269. Nagasawa, S., Shinohara, H., and Shiba, M.
1965. Chemosterilants of Insects VI. Sterilizing effect of Dowco-186 on the azuki bean weevil, *Callosobruchus chinensis L.* with special reference to the hatchability of eggs deposited by the created adults. *Bochu- Kagaku 30*: 91–95.

270. Nakao, Y., Yamaguchi, E., and Machida, I.
1962. A comparison of the mode of action between ethyleneimine and TEM in the inductions of lethal mutation and translocation. *Japan. J. Genet. 37*: 402.

271. Nasrat, G. E.
1965. Maleic hydrazide, a chemical mutagen in *Drosophila melanogaster. Nature 207*: 439.

272. ———.
1967. The effect of pretreatment with maleic hydrazide of the mutation rate induced by gamma radiation in *Drosophila melanogaster. Nippon Ideng. Zasshi 42*: 39–42.

273. Nethery, A. A., and Wilson, G. B.
1965. Classification of the cytological activity of phenols and aromatic organophosphates. *Cytologia 31*: 270–275.

274. ———.
1966. Classification of the cytological activity of phenols and aromatic organophosphates. *Cytologia 31(3)*: 270–275.

275. North, D. T.
1967. Sperm storage: Modifications of recovered dominant lethal mutations induced by Tetramine and

analogues. *Mutation Res. 4*: 225–228.

276. Northrop, J. H.
1963. The origin of bacterial viruses. VII. The effect of various mutagens (urethane, ethyl urethane, hydrogen peroxide, desoxycholate, maleic hydrazide, butadiene dioxide, Triethylenemelamine, Versene, and acriflavine) on the proportion of virus-producing and streptomycin-resistant cells in cultures of *B. megatherium* 20. *J. Gen. Physiol. 46*: 971–981.

277: Nygren, A.
1949. Cytological studies of the effects of 2, 4-D, MCPA, and 2,4,5,-T on *Allium cepa. Kgl. Lantbruks-Hoegskol Ann. Sweden 16*: 723–728.

278. Obe, G.
1968. Chemische konstitution and mutagene Wirkung. V. Vergleichende unterschung der Wirkung von Athyleniminen auf menschliche Leukozyten-chromasomen. *Mutation Res., 6(3)*: 467–471.

279. Ockley, C. H.
1957. A quantitative comparison between the cytotoxic effects produced by proflavin acetylethyleneimine and TEM on root tips of *Vicia faba. J. Genetics 55*: 525–550.

280. ———.
1960. Chromatid aberration induced by ethyleneimines. I. *Abh. dt. Akad. Wiss. Berlin* KL Med, *1*: 47–53.

281. Ofengand, J.
1965. A chemical method for the selective modification of pseudouridine in the presence of other nucleosides. *Biochem. and Biophys. Res. Comm. 18*: 192–201.

282. Olsen, P. J., Zaliks, S., Breakey, W. J., and Brown, D. A.
1951. Sensitivity of wheat and barley at different stages of growth to treatment with 2,4-D. *Agron J. 43*: 77–83.

283. Ostergren, G.
1944. Colchicine mitosis, chromosome contraction, narcosis, and protein chain folding. *Hereditas 30*: 429–467.

284. Ottoboni, A.
1969. Effect of DDT on reproduction in the rat. *Toxicol. Appl. Pharmacol. 14*: 74–81.

285. Palmquist, J., and LaChance, L. E.
1966. Comparative mutagenicity of two chemosterilants, Tepa and Hempa, in sperm of *Bracon hebetor. Science 154*: 915–917.

286. Pate, B. D., and Hays, R. L.
1968. Histological studies of testes in rats treated with certain insect chemosterilants. *J. Econ. Ent. 61*: 32–34.

287. Pickett, A. D., and Patterson, N. A.
1963. Arsenates: Effect on fecundity in some Diptera. *Science 140*: 493–494.

288. Poczatek, A.
1968. A case of congenital syndrome of locomotor

organ abnormalities. *Pediat. Pol.* 43: 487–488.

289. Prasad, I., and Pramer, D.
1968. Genetic effects of Ferbam on *Aspergillus niger* and *Allium cepa. Phytopathology 58*: 1188–1189.

290. ———.
1968. Mutagenic activity of some chloroanilines and chlorobenzenes. *Genetics 60*: 212–213.

291. ———.
1969. Cytogenetic effects of Propanil and its degraded products on *Allium cepa*. *Cytologia 34*: 351–352.

292. Quidet, P., and Hitier, H.
1948. Obtaining polyploid plants by treatment with hexachlorocyclohexane and polochlorocyclane sulphide. *Compt. Rend. Acad. Sci.* (Paris) *226*: 833–835.

293. Quinby, G. E.
1967. Physiochemical changes in pesticides after formulation causing health hazards. *Toxicol. Appl. Pharmacol. 10*: 390.

294. Raettig, H.
1955. Chemische Inacktivierung von Bakteriophagen durch Triathylenimin (TEM) *Zentr. Bakteriol. Parasitent. 163*: 245–247.

295. Raettig, H., and Uecker, W.
1955. Inaktivierung von Bakteriophagen durch Athyleniminderivate. *Naturwissenschaften 42*: 490.

296. Rai, K. S.
1964a. Cytogenetic effects of chemosterilants in mosquitoes. I. Apholate-induced aberrations in the somatic chromosomes of *Aedes aegypti L. Cytologia 29*: 346–353.

297. ———.
1964b. Cytogenetic effects of chemosterilants in mosquitoes. II. Mechanisms of Apholate-induced changes in fecundity and fertility of *Aedes aegypti* (L.) *Biol. Bull. 127*: 119–131.

298. Rake, A. V., and Tener, G. M.
1966. Effect of cyanoethylation of yeast transfer ribonucleic acid on its amino acid acceptor activity. *Biochem. 5(12)*: 3992–4003.

299. Ramel, C.
1967. Genetic effects of organic mercury compounds. *Acta Oecol. Scand.* (Suppl) *9*: 35–37.

300. ———.
1967. Genetic effects of organic mercury compounds. *Hereditas 57*: 445–447.

301. ———.
1969. Genetic effects of organic mercury compounds. I. Cytological investigations on Allium roots. II. Chromosome segregation in *D. melanogaster. Hereditas 61*: 208–254.

302. Rapoport, I. A.
1948a. Alkylation of gene molecule. *Dokl. Akad. Nauk SSSR 59*: 1183–1186.

303. ———.

1948b. Action of ethylene oxide glycidol, and glycols on gene mutations. *Dokl. Akad. Nauk SSSR 60*: 496–472.

304. ———.
1960. Reaction of gene proteins with ethylene chloride. *Dokl. Akad. Nauk SSSR 134*: 1214–1217.

305. ———.
1962. The interaction between the ethylenimine and the gene protein and hereditary variation. *Byul. Mosk. Obshch. Isp. Prirody. Otd. Biol. 67(1)*: 96–114.

306. ———.
1964. Mutagenic activity of Merthiolate and the mutagenic field. *Moskov. Obshch. Isp. Prirody. B. Otd. Biol. 69(5)*: 112–129.

307. ———.
1965. Mutational effect of paranitroacetophenylene-triphenylphosphine in connection with the additivity of mutational elementary components. *Dokl. Akad. Nauk SSSR 160(3)* 707–709.

308. Ratnayake, W.
1968. Effects of storage on dominant lethals induced by alkylating agents (Triethylene melamine and ethylenimine.) *Mutation Res. 5*: 271–278.

309. ———, Strachan, C., and Auerbach, C.
1967. Genetical analysis of the storage effect of Triethylene melamine (TEM) on chromosome breakage in *Drosophila. Mutation Res. 4*: 380–381.

310. Ristich, S. S., Ratcliffe, R. H., and Perlman, D.
1965. Chemosterilant properties, cytotoxicity, and mammalian toxicity of Apholate and other P–N ring chemicals. *J. Econ. Ent. 58*: 929–932.

311. Robens, J. F.
1969. Teratologic studies of Carbaryl, Diazinon, Norea, Disulfiram and Thiram in small laboratory animals. *Toxicol. Appl. Pharmacol. 15*: 152–163.

312. Robinson, A. G.
1961. Effects of Amitrole, Zytron, and other herbicides or plant growth regulators on the pea aphid. *Acyrthosiphon pisum (macrosiphon pisi)* caged on broad bean, *Vicia faba*. L. *Can. J. Plant Sci. 41*: 413–417.

313. Röhrborn, G.
1959. Mutation tests with melamine and Trimethylamelamine. *Dros Inf. Serv. 33*: 156.

314. ———.
1962. Chemische Konstitution and mutagene Wirkung. II. Triazinderviate. *Z. Vererb. 93*: 1–6.

315. Rose, F. L., Hendry, J. A., and Walpole, A. L.
1950. New cytotoxic agents with tumour inhibitory activity. *Nature 165*: 993–996.

316. Rosen, C. G., Ackerfors, H., and Nilsson, R.
1966. Organic mercury compounds as fungicides—economic requirements and health hazards. *Svensk. Kem. Tidskr 78*: 8–19.

317. Rowe, F. P., and Redfern, R.
1965. Toxicity tests on suspected Warfarin-resistant house mice (*Mus musculus* L.). *J. Hyg. 63*: 417–425

318. Rudenberg, L., Foley, G. E., and Winter, W. D.
1955. Chemical and biological studies on 1,2-dihydro-s-triazines. XI. Inhibition of root growth and its reversal by *Citrovorum* factor. *Science 121*:899–900.

319. Rybakova, M. N.
1966. On the toxic effect of Sevin on animals. *Gig. Sanit. 31*: 42–47.

320. Ryland, A. G.
1948. A cytological study of the effects of colchicine, indole-3-acetic acid, potassium cyanide, and 2,4-D on plant cells. *J. Elisha Mitchell Sci. Soc. 64*: 117–125.

321. Saha, N., and Bose, S. K.
1967. Chemical mutagenesis in *Serratia* phage kappa. *Indian J. Exp. Biol. 5*: 241–242.

322. Sass, J.
1937. Histological and cytological studies of Ethyl mercury phosphate poisoning in corn seedlings. *Phytopathology 27*: 95–99.

323. Sawamura, S.
1953. The effect of 2,4-Dichlorophenoxyacetic acid on the staminal hair cells of *Tradescantia* in vivo. I. 2,4-D sodium hydrate effect on the mitotic cells. *Bot. Mag. Tokoyo 66*: 155–160.

324. ———.
1955. Cytological studies on the effect of maleic hydrazide on the mitotic cells in vivo. *Bull. Utsunomiya Univ. 5*: 31–36.

325. ———.
1964. Cytological studies on the effect of herbicides on plant cells *in vivo*. I. Homonic Herbicides. *Cytologia 29*: 86–102

326. ———.
1965. Cytological studies of herbicides on plant cells in vivo. II. Non-hormonic herbicides. *Cytologia 30*: 325–348.

327. Sax, K., and Sax, H.
1968. Possible mutagenic hazards of some food additives, beverages, and insecticides. *Japan. J. Genetics 43*: 89–94.

328. Scholes, M. E.
1955. The effects of Aldrin, Dieldrin, Isodrin, Endrin, and DDT on mitosis in roots of the onion (*Allium cepa*). *J. Hort. Science 30*: 181–187.

329. Schwartz, P. H.
1965. Effects of Apholate, Metepa, and Tepa on reproductive tissues of *Hippelates pusio*. *J. Inv. Path. 7*: 148–151.

330. Scott, M. A., and Struckmeyer, B. E.
1955. Morphology and root anatomy of squash and cucumber seedlings treated with Isopropyl N-(3-chlorophenyl) carbamate. *Bot. Gaz. 117*: 37–45.

331. Sharma, A. K., and Ghosh, S.
1965. Chemical basis of the action of cresols and nitrophenols on chromosomes. *The Nucleus 9*: 183–190.

332. Shellenberger, I. E., Skinner, W. A., and Lee, J. M.
1967. Effect of organic compounds on reproductive processes. IV. Response of Japanese quail to alkylating agents. *Toxicol. Appl. Pharmacol. 10*: 69–78.

333. Shinohara, H., and Nagasawa, S.
1963. Chemosterilants of insects. I. Sterilizing effect of Apholate and METEPA on adults of the adzuki bean weevil. *Entomol. Expt. Appl. 6*: 263–267.

334. Sidorov, B. N., Sokolov, N. N., and Andreev, V. S.
1966. Highly active secondary alkylating mutagens. *Genetika 7*: 124–133.

335. Silberger, J., Jr., and Skoog, F.
1953. Changes induced by indoleacetic acid in nucleic acid contents and growth of tobacco pith tissue. *Science 118*: 443–444.

336. Simon, K.
1952. Untersuchung von Derivaten des Hydrazins auf ihre cytostatische Wirkung am Ascitestumor der Maus. *Z. Naturforsch. 7b*: 531–536.

337. Sjodin, J.
1962. Some observation in $X_1$ and $X_2$ of *Vicia faba* after treatment of *Vicia faba* with different mutagens. *Hereditas 48*: 565–586.

338. Skinner, W. A., Cory, M., Shellenberger, T. E., and De Graw, I. I.
1966. Effect of organic compounds on reproductive processes. 3. Alkylating agents derived from various diamines. *J. Med. Chem. 9*: 520–522.

339. ———.
1967. Effect of organic compounds on reproductive processes. V. Alkylating agents derived from aryl-, aralkyl-, and cyclohexyl-methyl enediamines. *J. Med. Chem. 10*: 120–121.

340. Slizynska, H.
1953. Cytological analysis of formaldehyde induced chromosomal changes in *Drosophila melanogaster*. *Proc. Royal Soc. Edinburgh 66*: 288–304.

341. ———.
1963. Mutagenic effect of x-rays and formaldehyde food in spermatogenesis of *Drosophila melanogaster*. *Genet. Res.* (*Camb.*) *4*: 248–257.

342. Smalley, H. E., Curtis, H. M., and Earl, F. L.
1968. Teratogenic action of Carbaryl in beagle dogs. *Tox. and Appl. Pharm. 13*: 392–403.

343. Smith, H. H., and Srb, A. M.
1951. Induction of mutation with $\beta$-propiolactone. *Science 114*: 490–492.

344. ———, and Lotfy, T. A.
1954. Comparative effects of certain chemicals on *Tradescantia* chromosomes as observed at pollen tube mitosis (Ethylene oxide). *Am. J. Bot. 41*: 589–593.

345. ———, and Lotfy, T. A.

(345. ———, and Lofty, T. A.)
1955. Effects of β-propiolactone and ceepryn on chromosomes of *Vicia* and *Allium*. *Am. J. Bot. 42*: 750–758.

346. ———, and Salkeld, E. H.
1966. The use and action of ovicides. *Ann. Rev. Ent. 11*: 331–368.

347. Sobels, F. H.
1956a. The effect of formaldehyde on the mutagenic action of x-rays in *Drosophila*. *Experientia 12*: 318.

348. ———.
1956b. Organic peroxides and mutagenic effects in Drosophila. *Nature 177*: 979–982.

349. Srivastava, L. M.
1966. Induction of mitotic abnormalities in certain genera of tribe *Vicieae* by Paradichlorobenzene. *Cytologia 31(2)*: 166–171.

350. Steinberger, E.
1962. A quantitative study of the effect of an alkylating agent (Triethylenemelamine) on the seminiferous epithelium of rats. *J. Reprod. Fertil. 3*: 250–259.

351. ———, Nelson, W. O., Boccabella, A., and Dixon, W. J.
1959. The radiomimetic effect of Triethylenemelamine on reproduction in the male rat. *Endocrinology 65*: 40–50.

352. Steinegger, E., and Levan, A.
1947. The effect of chloroform and colchicine on *Allium Hereditas 33*: 515–525.

353. Stroyev, V. S.
1968. Cytogenetic activity of the herbicides Simazine and maleic acid hydrazide. *Genetika 4(12)*: 130–134.

354. ———.
1968. The mutagenic effect by the action of herbicides on barley. *Genetika 4(11)*: 164–168.

355. Stumm-Tegethoff, B.
1964. Formaldehyd-Verunreini-gungen als mutagenes Prinzip bei *Drosophila melanogaster. Naturwissenschaften 51*: 646–647.

356. Sturtevant, F. M. Jr.
1953. Studies of the mutagenicity of phenol in *Drosophila melanogaster*. *J. Hered. 43*: 217–220.

357, Swanson, C. P., LaVelle, G. A., and Goodgal, S. H.
1949. Ovule abortion in *Tradescantia* as affected by aqueous solutions of 2, 4–D. *Am. J. Bot. 36*: 170–175.

358. ———, and Merz, T.
1959. Factors influencing the effect of β-propiolactone on chromosomes in *Vicia faba*. *Science 129*: 1364–1365.

359. Tatum, E. L.
1947. Chemically induced mutations and their bearing on carcinogenesis. *Ann. N. Y. Acad. Sci. 49*: 87–97.

360. ———, Barrett, R. W., Fries, N., and Bonner, D.
1950. Biochemical mutant strains of *Neurospora* produced by physical and chemical treatment. *Am. J, Botany 37*: 38–46.

361. Templeman, W. G., and Sexton, W.
1946. The differential effect of synthetic plant growth substances and other compounds upon plant species. I. Seed germination and early growth responses to alphanaphthylacetic acid and compounds of the general formula aryl-$OCH_2COOR$. *Proc. Royal Soc. B. 133*: 300–313.

362. ———.
1945. Effect of some arylcarbamic esters and related compounds upon cereals and other plant species. *Nature 156*: 630.

363. Tenhet, J. N.
1947. Effect of sublethal dosages of Pyrethrum on oviposition of the cigarette beetle. *J. Econ. Ent. 40*: 910–911.

364. Terranova, A. C., and Schmidt, C. H.
1967. Purification and analysis of HEMPA by chromatographic techniques. *J. Econ. Ent. 60*: 1659–1663.

365. Toropova, G. P., and Egorova, E. V.
1967. Changes in the nucleic acid content in the tissues and nuclei of the white rat liver after intake of Chlorophus (*Dipterex*). *Vop. Pitan. 26*: 16–21.

366. Tubaro, E., and Bulgini, M. J.
1968. Cytotoxic and antifungal agents: their body distribution and tissue affinity. *Nature 218*: 395–396.

367. Uecker, W.
1963. Inaktivierung von Mikrooganismen durch alkylierende Verbindungen. IV. Reaktivierung von Bakteriophagen nach Inaktivierung mit Trimethylolymelamine. *Zentr. Bakt. Parasit. 189*: 178–188.

368. Underhill, J. C., and Merrell, D. J.
1966. Fecundity, fertility, and longevity of DDT-resistant and susceptible populations of *Drosophila melanogaster*. *Ecology 47*: 140–142.

369. Unrau, J.
1952–1953. Cytogenetic effects of 2,4–D on cereals. *Can. Seed Growers Assoc.* (annual report): 25–28.

370. ———.
1953–1954. Cytogenetic effects of 2,4–D on cereals. *Can. Seed Growers Assoc.* (annual report): 37–39.

371. ———, and Corns, W. G.
1950. Cytological and physiological effects of 2,4–D applied to cereal grains at different stages of growth. *Research Report*, 7th Annual Northcentral Weed Conference: 279.

372. ———, and Larter, E. N.
1952. Cytogenetical responses of cereals to 2,4–D. I. A study of meiosis of plants treated at various stages of growth. *Can. J. Botany 30*: 22–27.

373. Vaarama, A.
1947. The influence of DDT pesticides upon plant mitosis. *Hereditas 33*: 191–219.

374. Vandenbergh, J. G., and Davis, D. E.
1962. Gametocidal effects of Triethylenemelamine on redwinged blackbirds. *J. Wildlife Man. 26*: 366.

375. Veleminsky, J., and Gichner, T.
1963. Cytological and genetic effects of the insecticide Systox on *Vicia faba* L. and *Arabidopsis thaliana* L. *Biol. Plantarum 5*: 41–52.

376. Verin, V. K.
1968. Atypical mitosis in the rat liver under intoxication with $CCl_4$. *Arkh. Anat. 55*: 63–66.

377. Vogt, M.
1950. Analyse durch Athylurethan bei *Drosophila* induzierter Mutationen. *Z. Ind. Abst. Vererb. 83*: 324–340.

378. Voogd, C. E., and Vet, P. V. D.
1969. Mutagenic action of ethylene halogenhydrins. *Experientia 25*: 85–86.

379. Wagner, R. P., Maddox, C. H., Fuerst, R., and Stone, W. S.
1950. The effect of irradiated medium, cyanide, and peroxide on the mutation rate in *Neurospora*. *Genetics 35*: 237–248.

380. Walker, A. I. T., Neill, G. H., Stevenson, E. E., and Robinson, J.
1969. The toxicity of Dieldrin (HEOD) to Japanese quail. *Toxicol. Appl. Pharmacol. 15*: 69–73.

381. Walpole, A. L., Roberts, D. C., Rose, F. L., Hendry, J. A., and Homer, R. F.
1954. Cytotoxic agents: IV. The carcinogenic actions of some monofunctional ethyleneimine derivatives. *Brit. J. Pharm. 9*: 306–323.

382. Ware, G. W., and Good, E. E.
1967. Effects of insecticides on reproduction in the laboratory mouse. IV. Mirex, Telodrin and DDT. *Abst. 6th Ann. Mtg. Toxicol. Appl.* Atlanta, Ga.

383. Weidhaus, D. E., Ford, H. R., Graham, J. B., and Smith, C. N.
1961. Preliminary observations on chemosterilization of mosquitos. *N.J. Mosq. Extermin. Ass. Proc. 48*: 106–109.

384. Weihe, M.
1967. Effects of DDT on reproduction in hens. *Acta Pharmacol. et Toxicol.* (Kobenhavn) *25* (Suppl. 4): 54.

385. West, S. H., Hanson, J. B., and Key, J. L.
1960. Effect of 2,4–D on nucleic acid and protein content of seedling tissue. *Weeds 8*: 333.

386. Westergaard, M.
1955. Chemical mutagenesis in relation to the concept of the gene. *Experientia 13*: 224–234.

387. Williams, R. T.
1969. The fate of foreign compounds in man and animals. *Pure and App. Chemistry 18*: 129–141.

388. Wills, J. H., Jameson, E., Stein, A., Serrone, D., and Coulston, F.
1967. Effects of oral doses of Carbaryl on man. *Abst. 6th ann. Mtg. Toxicol. Appl.* Atlanta, Ga., Mar. 23–25.

389. Wilson, S. M., Daniel, A., and Wilson, G. B.
1956. Cytological and genetic effects of the defoliant

Endothall. *J. Heredity 47*: 151–155.

390. Wolfe, H. R., Durham, W. F., and Armstront, J. F. 1963. Health hazards of the pesticides Endrin and Dieldrin. *Arch. Environ. Health 6*: 458–464.

391. Wright, G. J. 1967. Tris(1-aziridinyl)phosphine oxide: studies on distribution and metabolism in the rat using Carbon-14. *Toxicol. Appl. Pharmacol. 10*: 400.

392. ———, and Roe, V. K. 1967. Ethyleneimine: Studies of the distribution and metabolism in the rat using Carbon-14. *Toxicol. Appl. Pharmacol. 10*: 400.

393. Wuu, K. D., and Grant, W. F. 1966. Induced abnormal meiotic behavior in a barley plant (*Hordeum vulgare* L.) with the herbicide Lorox. *Phyton 23(1)*: 63–67.

394. ———. 1966. Morphological and somatic chromosomal aberrations induced by pesticides in barley (*Hordeum vulgare*). *Can. J. Genet. Cytol. 8*: 481–501.

395. ———. 1967. Chromosomal aberrations induced by pesticides in meiotic cells of barley. *Cytologia 32*: 31–41.

396. ———. 1967. Chromosomal aberrations induced in somatic cells of *Vicia faba* by pesticides. *Nucleus Int. J. Cytol. Allied Top. 10(1)*: 37–46.

397. ———. 1967. Chromosomal aberrations induced by a plant growth retarding chemical in barley. *Bot. Bull. Acad. Sinia (Japan) 8*: 191–198.

398. Yoshida, M., and Ukita, C. 1965. Selective modifications of inosine and $\phi$-Uridine with acryolnitrile out of the other ribonucleosides. *J. Biochem. 57*: 818–821.

399. ———. 1965. Reaction rates of acrylonitrile and with $\phi$-Uridine residues in transfer ribonucleic acid. *J. Biochem. 58*: 191–193.

400. Yule, W. N., Hoffman, I., and Parups, E. V. 1966. Insect toxicological studies of maleic hydrazide translocated in the potato plant. *Bull. Environ. Contam. Toxicol. 1*: 251–256.

401. ———, Parups, E. V., and Hoffman, I. 1966. Toxicology of plant-translocated maleic hydrazide. Lack of effects on insect reproduction. *J. Agr. Food Chem. 14*: 407–409.

402. Zacharias, M., and Ehrenberg, L. 1962. Induction of leaf spots in leguminous plants by nucleotoxic agents, I. *Hereditas 48*: 284–306.

403. Zavon, M. R., Tye, R., and Stemmer, K. 1967. Effect of Dieldrin insecticides on Rhesus monkeys after three years of ingestion. *Abstr. 6th Ann. Mtg. Toxicol. Soc.* Atlanta, Ga. Mar. 23–25.

404. Zuckel, J. W. 1952–1955. *Literature summary on maleic hydrazide.* Naugatuck, Conn.: U.S. Rubber Co. Publications.

# Index

See also pp. 70–175 for alphabetical tabulation and cross index of pesticides.

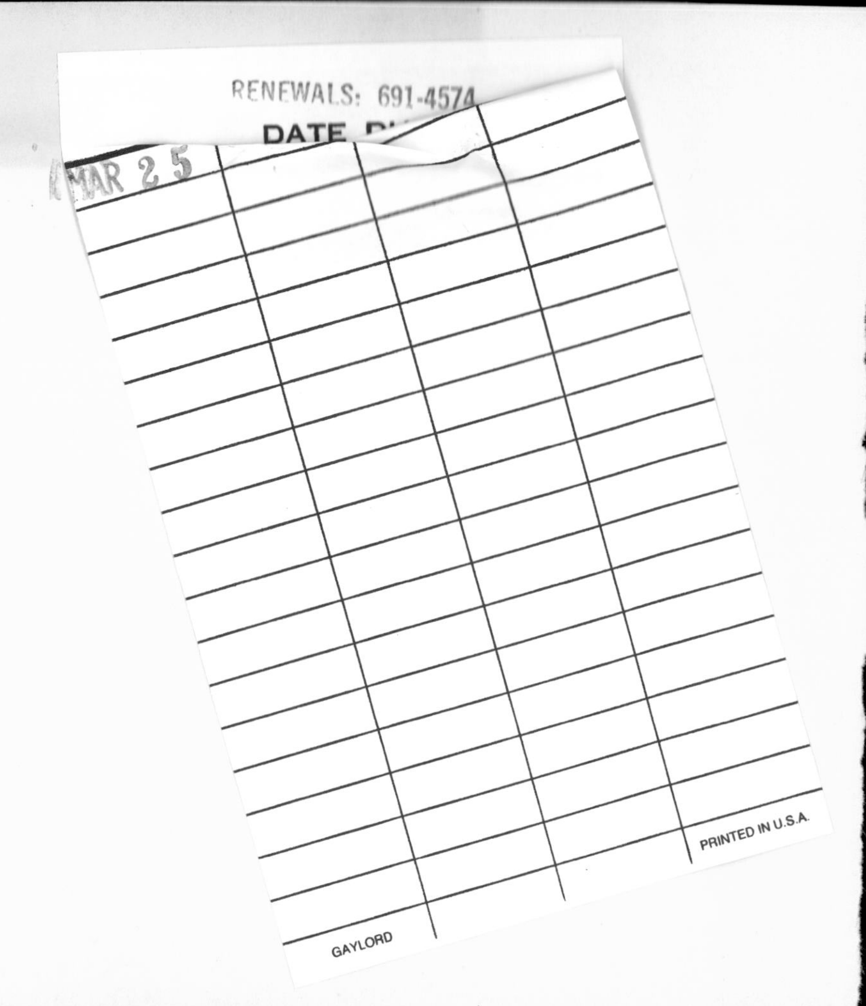
RENEWALS: 691-4574
DATE
MAR 25
GAYLORD
PRINTED IN U.S.A.